D0895283

Humid Landforms

AN INTRODUCTION TO SYSTEMATIC GEOMORPHOLOGY

The following volumes have been published

An Introduction to Systematic Geomorphology
Volume One

Humid Landforms
Ian Douglas

The MIT Press
Cambridge, Massachusetts

GB
446
D6

First MIT Press edition, 1977
First published in Australia 1977

Printed in Singapore for the Australian National University Press,
Canberra

© Ian Douglas 1977

This book is copyright. Apart from any fair dealing for the purpose
of private study, research, criticism, or review, as permitted under
the Copyright Act, no part may be reproduced by any process
without written permission. Inquiries should be made to the
publisher.

Library of Congress catalog card number: 77–76682
ISBN 0–262–04054–9

INTRODUCTION TO THE SERIES

This series is conceived as a systematic geomorphology at university level. It will have a role also in high school education and it is hoped the books will appeal as well to many in the community at large who find an interest in the why and wherefore of the natural scenery around them.

The point of view adopted by the authors is that the central themes of geomorphology are the characterisation, origin, and evolution of landforms. The study of processes that make landscapes is properly a part of geomorphology, but within the present framework process will be dealt with only in so far as it elucidates the nature and history of the landforms under discussion. Certain other fields such as submarine geomorphology and a survey of general principles and methods are also not covered in the volumes as yet planned. Some knowledge of the elements of geology is presumed.

Four volumes will approach landforms as parts of systems in which the interacting processes are almost completely motored by solar energy. In humid climates (Volume One) rivers dominate the systems. Fluvial action, operating differently in some ways, is largely responsible for the landscapes of deserts and savannas also (Volume Two), though winds can become preponderant in some deserts. In cold climates, snow, glacier ice, and ground ice come to the fore in morphogenesis (Volume Three). On coasts (Volume Four), waves, currents, and wind are the prime agents in the complex of processes fashioning the edge of the land.

Three further volumes will consider the parts played passively by the attributes of the earth's crust and actively by processes deriving energy from its interior. Under structural landforms (Volume Five), features immediately consequent on earth movements and those resulting from tectonic and lithologic guidance of denudation are considered. Landforms directly the product of volcanic activity and those created by erosion working on volcanic materials are sufficiently distinctive to warrant separate treatment (Volume Six). Though karst is undoubtedly delimited lithologically, it is fashioned by a special combination of processes centred on

v

83723

solution so that the seventh volume partakes also of the character of the first group of volumes.

J. N. Jennings
General Editor

PREFACE

Landforms created by running water dominate the land surface of earth. However, although the role of water is seen everywhere, it is seen at its best in those regions where the climates are wet enough to support a forest vegetation with a continuous canopy. The seasonal fluctuations in the character of precipitation with snow in winter and rain in summer which characterise cool temperate forest climates and the legacies of recent past cold periods in high latitudes means that the landforms of the humid tropics should be regarded as the 'normal' or 'type' features due to erosion by running water.

This discussion of humid landforms therefore attempts to break with traditional approaches and to discuss humid landforms from the standpoint of the humid tropics. In addition, it seeks to demonstrate that the processes creating and destroying landforms are also those that regulate biotic activity at the earth's surface. The approach followed in this book is to describe the processes affecting the evolution of landforms in terms of the circulations of energy, water and materials before introducing the complication of legacies of different ages from the past. Theories of landform evolution are briefly reviewed in the final chapter. It is hoped that this arrangement will encourage readers to think about the variety of ways in which landforms are changing at present and to use their own experience to evaluate the theories currently discussed in geomorphology.

Armidale
1976

I. D.

ACKNOWLEDGMENTS

The encounters with humid landforms that have provided the experience from which this book stems have been fostered by research opportunities provided by the Universities of Oxford, Strasbourg, Hull, Malaya and New England together with the Australian National University. The various scholarships and research grants which I held at those institutions made travel to tropical and temperate environments possible, and fieldwork under the guidance of Marjorie Sweeting, Jean Tricart, Joe Jennings and with the advice of the late Robert Ho greatly enhanced my understanding of the landforms of those areas. A special debt is owed to Maureen Douglas, Ronald Douglas, Ken Gregory and Anton Imeson, who all commented on the text and to Tony Spenceley who helped compile some of the diagrams. Denis Dwyer and Ed Derbyshire provided ideal conditions at Keele University during study leave for the final revision of the book. Cartography by Reg Dean, Keith Scurr, Wendy Wilkinson, Johnnie Ngai, Eric Schofield, Tony King, Bill Neal, Rudi Boskovic, Paul Branscheid, Mick Roach and Mrs A. Patrick improved the diagrams, while Margaret Watson, Sue Nano and Sue Davis typed successive drafts of the manuscript. The Director of National Mapping, Department of National Resources, Australia; the Commonwealth Scientific and Industrial Research Organisation, Australia; the Australian Information Service and Joe Jennings have generously supplied photographs. The University of Exeter and Dr Ken Gregory gave permission for the reproduction of figure 31. However, above all thanks are particularly due to Joe Jennings, editor of this series, whose rigorous criticism has constantly helped to improve the quality of the text, and to the editorial staff of the A.N.U. Press, for their patience and constructive criticism.

CONTENTS

FIGURES

PLATES

TABLES

I

HUMID LANDFORMS
AND DENUDATION SYSTEMS

Water is important in the evolution of almost all landforms, for occasional water movement is significant even in desert environments or beneath ice sheets. As water is one of the main agencies transporting detritus and solutes from one area to another, a study of the hydrological cycle and its associated sections of the geochemical cycle is essential for understanding the ways in which landforms evolve. Nevertheless, processes associated with water movement are the dominant landforming agents over only part of the earth's surface. Over much of the earth, there are one or more periods of the year with a water shortage, when evapotranspiration uses nearly all available and stored water leaving little for runoff to rivers and thus for transporting detritus or solutes.

In humid areas periods of water deficiency are so short as to permit the growth of a forest vegetation, the maintenance of permanent rivers and a general downward movement of moisture in the soil. Forest vegetation is significant in landform development because any organic matter covering rocks regulates the action of water on those rocks. Thus, in this book, the milieu of humid landform evolution is taken to be a forest environment where the canopy of one tree touches that of the next, thus protecting the soil from the sun and reducing the direct impact of raindrops on the soil. Many areas which could support a forest vegetation do not do so now, as the forests have been cleared, often by systematic burning, as in the highlands of Papua New Guinea, or by modern machinery, as in clearfelling operations in equatorial lowland rain forests. The landforms of such areas, however, retain, in all but the smallest details, the characteristics of humid landforms.

The forested areas of the world fall into two main groups, the tropical rain forests and the forests of temperate lands, although there are, or were before clearance, virtually continuous forests from the tropics to cool temperate latitudes on the eastern sides of the continental land masses. Erosion and aggradation in these forest zones is accomplished by rainfall and running water, although in the cooler parts of the temperate forests snow lies on the ground

1

for part of the year. The boreal forest areas affected by frozen ground conditions have landforms typical of cold climates (Davies, 1969), which are beyond the scope of this work.

These forest environments range from the maritime temperate forests of Europe and North America, where a single tree species may dominate the forest formation, to the rich complex ecosystem of the tropical rain forest. This botanical diversity is associated with a geomorphic diversity, since climatic differences between the thermal seasons of the temperate forest zone and the moisture seasons of the tropical forest zone produce variations in the intensity and rate of operation of processes and in the relative importance of various landforming processes. Nevertheless on the global scale there is essentially one pattern of humid landform development, that of the forest fluvial landscape pattern best developed in the humid tropics. The ideal types of humid landforms should be sought in the humid tropics, which are far less affected by the legacies of past climatic changes than temperate lands and which escape such seasonal changes in the form of precipitation as winter snowfall.

Seasonal contrasts in runoff and erosion are particularly likely to occur when winter ground cover is less than in summer, and when lower winter evaporation rates increase rain or snow storm runoff (Imeson, 1970). Where snow accumulates during the winter, the spring thaw is the period of high runoff and erosion. Most erosion in the headwater channels of streams draining clay-rich sandstones in the Paris Basin of France occurs during a few weeks of spring thaw (Cailleux, 1948).

That humid tropical environments with minimal seasonal contrasts may provide basic information on fluvially eroded landscapes has long been recognised. When many students of landforms in the mid-nineteenth century were still thinking that valleys were cut by the action of the sea, Dana (1850) used the tropics of South America and the Pacific to demonstrate the true importance of 'gradual wear from running water derived from rains — gradual decomposition through the agency of the elements and growing vegetation'. His descriptions of valley formation in the Hawaiian islands and the observations of Oldham and other British workers in India, especially in the rain forests of Assam (Chorley, Dunn and Beckinsale, 1964: 386), greatly increased awareness of the significance of fluvial processes in landform development.

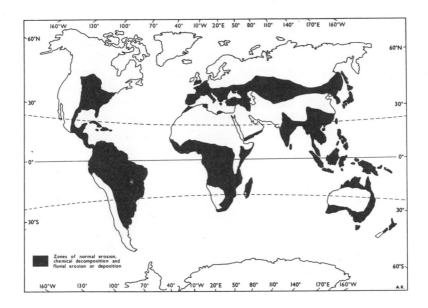

1   *Areas of normal erosion according to de Martonne*

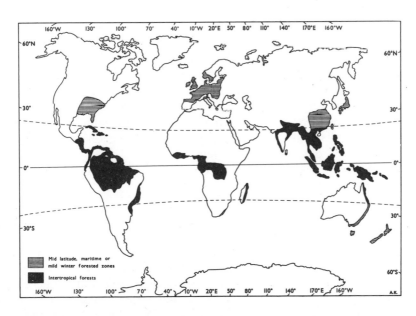

2   *Humid morphoclimatic regions according to Tricart and Cailleux*

Later, writers such as Behrmann (1921) and Sapper (1935) showed that the rain forests of the humid tropics were a special kind of morphogenic environment, where chemical and biotic processes were more powerful than in temperate forest environments. However, both forest environments were regarded as essentially part of the same fluvial system. In a broad review of climatic influences on landforms, Behrmann (1933: 426) pointed out that the processes operating in temperate forested regions resemble those of tropical forests, but operate on a different scale and at different rates. This notion is echoed in Garner's discussion of climatic geomorphology (1968), which emphasises the importance of ground cover.

The broad similarity of landform development in both tropical and temperate humid areas is recognised by the classification of all such areas as districts of normal erosion on de Martonne's map of dominant erosion forms (de Martonne, 1951) (Fig. 1). Nevertheless, de Martonne (1940, 1946, 1951) indicates that tropical and temperate humid landforms may differ in detail, some tropical landforms not having temperate counterparts. Tricart and Cailleux (1965) recognised distinct tropical and temperate humid morphoclimatic regions (Fig. 2).

Statements on the similarities and on the contrasts between temperate and tropical humid landforms may be reconciled by examining the scale appropriate to each statement. Convenient frames of reference are provided by the classification of geomorphic phenomena of Tricart (1965a) (Table 1) and the geographical scale of Haggett, Chorley and Stoddart (1965) (Table 2). The study of landforms involves the whole range of magnitudes down to Tricart's seventh order and to G-scale 11 to 12. However, Garner's threefold division of ground cover (1968) into:

1. effectively continuous plant cover in the intermediate and low latitudes;
2. ice cover in the higher latitudes and altitudes;
3. effectively exposed land (discontinuous plant cover) at any latitude;

is essentially a world scale division at Tricart's first order of magnitude and at a G-scale value of less than 2. From the tables, it is apparent that the distinction between tropical and temperate humid landforms operates at the second order of magnitude, or G-scale values of 2 to 3. In the middle order of magnitude,

TABLE 1 **Taxonomic classification of geomorphic phenomena (Tricart. 1965a)**

ORDER	SURFACE UNIT SIZE km²	CHARACTERISTIC OF UNITS, EXAMPLE	CORRESPONDING CLIMATIC UNITS	DOMINANT MORPHOGENETIC PROCESSES	TIME SPAN THROUGH WHICH FEATURES REMAIN (YRS)
1	10^7	Continents, ocean basins (major features of earth's surface)	Major zonal divisions affected by astronomical factors	Differences in crustal layers, sial and sima	10^9
2	10^6	Major structural zones (Scandinavian shield, Congo basin)	Major climatic types (inter-action of astronomical and geographical features)	Crustal movements, such as formation of geosynclines; influence of climate on dissection	10^8
3	10^4	Major structural units (Sydney basin, Darling Downs, highlands of Sri Lanka)	Variation with climatic types but without marked effects on landscape dissection	Tectonic units having close links with palaeogeographic conditions; rate of dissection influenced by lithology	
4	10^2	Simple tectonic features: mountain massifs, horsts, rift valleys (Bellenden Ker-Bartle Frere massif, Mulgrave-Russell corridor, Cumberland Plain)	Regional climates influenced by geographical features, especially in mountains	Predominant influence of tectonics with subsidiary lithologic effects	10^7
			— Limit of isostatic compensation —		
5	10	Tectonic accidents: anti-clines, monoclines	Local climates, influenced by orientation and character of the relief: adret, ubac, altitudinal level	Predominant influence of lithology and isostatic adjustment	10^6-10^7
6	10^{-2}	Relief forms: crest, terrace, cirque, moraine, alluvial fan, floodplain	Mesoclimate directly related to landform	Predominance of the morphodynamic factor, influenced by lithology	10^4
7	10^{-6}	Micro-features: solifluction lobes, gullies, terracettes	Microclimate directly linked to landform by autocatalysis	Predominance of the morpho-dynamic factor, influenced by lithology (e.g. *Silikaikarren* described in Chapter III)	10^2
8	10^{-8}	Microscopic features: details of solution or polishing etc.	Microenvironment	Influence of dynamics of environment and micro-structure of the rock	

Humid Landforms

TABLE 2 **G-Scale of Haggett, Chorley and Stoddart (1965)**

An absolute scale based on the surface area of the earth (Ga) which is successively divided by power of 10. This scale, by using the negative logarithms of areas compared with a unitary value for Ga, is able to accommodate the smallest features, even microscopic clay-sized fragments.

G-scale value

0	$= Ga$	$= 5.098 \times 10^8$
1	$= Ga \times 10^{-1}$	$= 5.098 \times 10^7$
2	$= Ga \times 10^{-2}$	$= 5.098 \times 10^6$
3	$= Ga \times 10^{-3}$	$= 5.098 \times 10^5$
4	$= Ga \times 10^{-4}$	$= 5.098 \times 10^4$
5	$= Ga \times 10^{-5}$	$= 5.098 \times 10^3$
6	$= Ga \times 10^{-6}$	$= 5.098 \times 10^2$
7	$= Ga \times 10^{-7}$	$= 5.098 \times 10^1$
8	$= Ga \times 10^{-8}$	$= 5.098$
12	$= Ga \times 10^{-n}$	$= 5.098 \times 10^n$

To calculate G for any area (Ra) being investigated:

$$G = \log Ca - \log Ra$$

In km^2 $= 8.7074 - \log Ra \ (km^2)$

In ha $= 10.7074 - \log Ra \ (ha)$

Examples of landform areas of various G-scale values

	G-scale value	Tricart Order
Pacific Ocean	0.54	1
Australia and Tasmania	1.81	1
China Sea	2.19	2
Great Artesian Basin	2.55	2
Nullarbor Plain	3.48	3
Basalt area of Atherton Tableland	5.30	4
Younger alluvium of Burdekin delta	5.64	4
Barron River catchment above crater	7.63	5
Lake Barrine explosion crater	8.36	6
Plan area of sand grain in Barron River at crater	13.31	8

the geomorphic effects of climate are not recognisable, for this is where tectonic and lithologic factors are dominant, in some instances giving rise to similar landforms in widely diverse morpho-climatic environments.

Figure 3 shows some of the complicated links in the geological and environmental controls of the evolution of landforms.

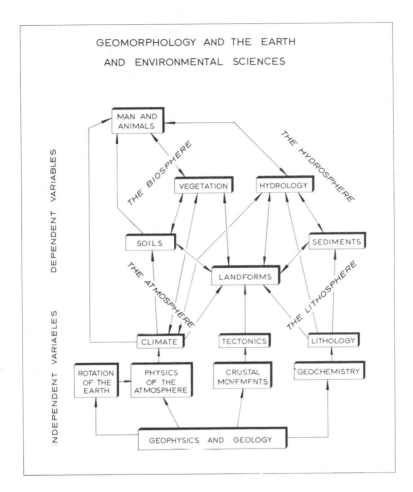

3　*Some geological and environmental controls of landform evolution*

Dynamic aspects of landform study

On different timescales, landforms may be considered either as virtually permanent or as part of the dynamic ever-changing pattern of landscape evolution. The appearance of the landsurface changes continually, even though the rate of change may be too slow for the human eye to perceive it. At other times, changes may be drastic, as when floodwaters cut new river channels, or landslips cause massive movements of earth down slopes. These changes are closely linked to the biological and climatic events affecting the earth's surface.

Every time raindrops hit the ground, some material is loosened, displaced, carried downhill or taken into solution. Movement of material is also achieved through organic activity, particularly in the soil. Plants, fungi, bacteria and other micro-organisms all play vital roles in the creation of the soil, and in humid regions the soil is the vital surface zone on which the atmospheric agencies work to change the landscape. Landforms thus have to be examined in terms of the complex of factors operating at the atmosphere-lithosphere interface, that vital life zone of the earth where plants and animals live.

Superimposed on these natural changes are those introduced by man. Changes due to major engineering works, such as major dams and hydroelectric schemes, the building of major roads, the open-cast or underground extraction of coal and other minerals, the artificial drainage of swamps and urban development can cause far more rapid changes in the landscape than would occur naturally (Sherlock, 1922; Jennings, 1965; Brown, 1970). Both these complex man-made changes and the less obvious indirect effects of human activity on natural geomorphic processes, such as the pollution of air and water and the modification of climatic and river regimes, should be considered when studying present-day landforms.

THE GEOMORPHOLOGICAL SIGNIFICANCE OF THE CIRCULATIONS OF ENERGY, WATER AND THE CHEMICAL ELEMENTS

The factors that affect landform evolution tend to interact to produce a morphogenic environment, which may be conveniently termed a denudation system (Douglas, 1969). In humid areas, as mentioned earlier, the operation of the denudation system is greatly influenced by biological activity in the plant community. This biological activity is in turn dependent on the inputs of energy, water, and chemical elements into the system. The biologists discuss such inputs and their consequences in terms of ecosystem dynamics, and this concept is very close to the geomorphological concept of erosion systems (Cholley, 1950) or denudation systems. Ecosystems are open systems in which energy inputs carry out work (Stoddart, 1965), leading to the storing of energy in plant communities. Energy capture by the fastest growing forest ecosystems at their most productive period possibly approaches the maximum possible under natural conditions (Ovington, 1962). In the open denudation system, energy works to achieve an adjustment between the various processes operating in the drainage network and slope system until eventually the rate of change is at a minimum consonant with the energy forces operating on the drainage basin. Such adjustment is incomplete in all landscapes because a whole series of environmental constraints prevents the attainment of a steady state of operation, or uniform rate of down-cutting throughout the drainage network.

A river provides an example of this adjustment towards equilibrium in a denudation system. During a flood such adjustments are rapid. Because each section of a stream is called upon to carry a greater discharge than it normally does, the velocity of flow increases greatly. The equilibrium state of a stream that has increased discharge is an increased cross-sectional water area and a higher velocity. But at the beginning of a flood, the stream is not in equilibrium with its new discharge: its area is too small and its velocity of flow is too great. The high velocity of flow enables the water to scour the bed and banks of the stream, which enlarges cross-sectional area. If the flood persists for some time, and if the bed of the stream is easily eroded, the stream will scour its channel until it reaches equilibrium under the conditions of increased discharge. As the water level drops and discharge falls, the stream is no longer in equilibrium with the enlarged channel. The reduced discharge cannot carry all the sediment in the channel and so the stream bed is gradually built up by deposited sediment until the stream is at equilibrium with a shallower channel.

Such scour and fill cannot take place readily when the stream passes over rock outcrops and in such circumstances either velocity increases sharply during floods, producing rapids, or if the gradient of the stream is slight, the rock bar will cause the water eventually to rise above the channel banks and inundate the surrounding countryside.

Over longer term periods, the nature of the rocks traversed by a river may affect the evolution of the channel, the drainage network and the related ground slopes. Examples of pronounced structural features affecting river patterns are the Cadell Fault Block on the Murray River in Australia, and the Iron Gates on the Danube. Such outcrops remain as major irregularities in a channel, despite the trend towards an equilibrium condition.

Energy flow in denudation systems

The ultimate source of practically all energy exchanged in ecosystems is the sun, annual incoming solar radiation being greatest in the high-pressure zones where the subtropical deserts occur, with a maximum over the eastern Sahara and Saudi Arabia. In the equatorial regions incoming radiation is reduced by cloud cover, while in the seasonally wet tropics it varies through the year (Fig. 4). Lower values of incoming radiation are also found in the humid temperature areas where cyclonic weather systems are again associated with considerable cloudiness.

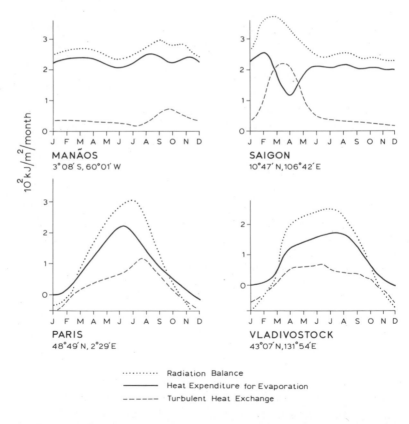

4 *Energy exchanges at tropical and temperate stations (after Budyko, 1958)*

Budyko (1958) shows that the heat balance has great significance for vegetation productivity, soil zonality and possibly also for the effectiveness of geomorphological processes. The geomorphological significance of the close relationship between the heat balance and the water balance arises because the processes changing landforms operate only because of a flow of energy from a higher potential or intensity to one that is lower, and thus the transformation of energy is the common denominator of all geomorphological processes.

One of the important systems powered by this energy flow is the hydrological cycle (Fig. 5), the evaporation phase of which uses up much of the heat available at the earth's surface (Fig. 4). As with other geomorphological phenomena, the hydrological cycle may be studied at different scales. On the world basis a global water balance may be established in which the equation $P = E + r$ (precipitation = evaporation + runoff) is applied to the oceans and the continents. For landforms the annual water balance of the land surface is of greatest interest. Budyko (1965) shows that the water balance involves approximately 730 mm precipitation of which 479 mm are evaporated, the remaining 251 mm being divided between

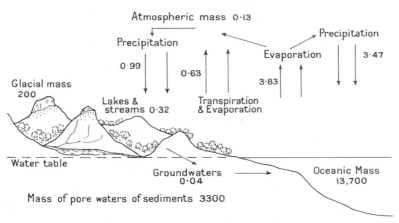

5   *The hydrological cycle showing the annual transfer of water, as well as the total mass of water in the oceans, in the pore waters of sediments, in ice, and in the atmosphere. Quantities are expressed in units of 10^{20} g. (Values are from Garrels and Mackenzie, 1971)*

170 mm lost as surface or flood runoff, and 81 mm as underground runoff. Soil moisture storage on an annual basis is estimated at 560 mm, much of this being returned to the atmosphere through the transpiration of plants, which accounts for at least 30 to 40 per cent of the water evaporated from the land. Soil moisture availability is one factor determining the rate of weathering, while most of the removal of debris is done by flood runoff.

The geochemical cycle

Study of the hydrological cycle on the river basin scale reveals the many different routes which may be taken by water from the atmosphere to rivers or before return to the atmosphere by evapotranspiration. All these lines of water movement, from the top of the vegetation canopy to the groundwater body, are links in the pattern of translocation of chemical elements and thus part of the geochemical cycle. The geochemical cycle represents the source materials section of Keller's equation (1954):

$$\text{source materials} + \text{energy} = \text{sedimentary rocks}$$

in which sedimentary rocks may be regarded as the depositional phase of the denudation system with which landform students are concerned. In the lithosphere, the geochemical cycle (Fig. 6) begins with the initial crystallisation of a magma; proceeds through the alteration and weathering of the igneous rock and the transportation and deposition of the material thus produced; continues through the compaction and other aspects of lithification or diagenesis of these deposits to metamorphism of successively higher grade until, eventually, by anatexis or palingenesis, magma is regenerated. In the study of landforms, the aspects of this geochemical cycle that are most significant are the phases of destruction of igneous, metamorphic and sedimentary rocks and the translocation of the chemical elements derived from them until their redeposition. The translocation of elements involves all the varied types of transfer of chemical elements through the lithosphere, biosphere, atmosphere and hydrosphere, including the exchanges of nutrients between living organisms and their environment (the biogeochemical cycles of Odum, 1963:53).

Increasingly, microbiological activity is being shown to be significant in landform development (Krumbein, 1969; Cailleux, 1969), the role of microflora varying with ecological conditions.

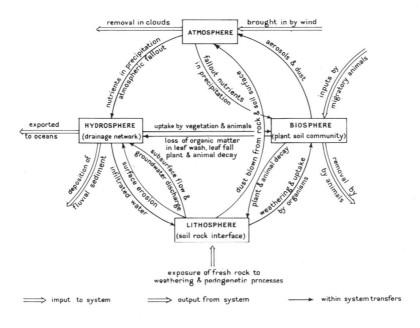

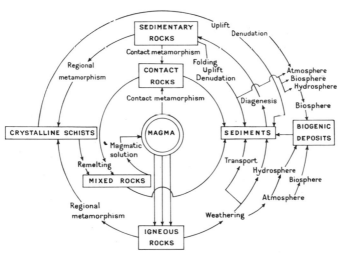

6 *The geochemical cycle. The upper diagram shows the exchanges of particular relevance to geomorphic processes, while the lower diagram (after Hinz, 1970) shows the complete cycle of geochemical changes affecting the earth's crust.*

For example, lichens in humid climates usually protect rocks from weathering, while the same organisms are one of the most effective agents of rock weathering under arid climatic conditions.

The denudation system is thus the product of the hydrological and geochemical cycles in a particular river basin or topographic unit, and depends ultimately on the manner in which the energy balance affects the seasonal and long term operation of those two fundamental circulations of water and rock materials. The study of landforms is thus in essence the analysis of the way in which fundamental physical and chemical processes operate, largely under the influence of local biological conditions, to produce various types of topographic features.

LANDFORM DESCRIPTION AND CLASSIFICATION

The inhabitants of the rugged selva landscape of New Guinea or the mountainous terrain of the humid regions of western North America will have different conceptions of a steep slope or high hill from those of the plains dwellers in the Mekong delta or Amazon lowland. Slight changes in elevation of the alluvial lowland which make the difference between being flooded and remaining dry may not be noticed by the mountain dweller. To avoid misunderstandings arising from contrasts in environmental perception between observers from different backgrounds, the scientific study of landforms must employ a standard series of descriptive methods.

Much information on humid landforms is derived from the analysis of maps, aerial photographs and other imagery obtained by remote sensing techniques. The methods of obtaining information on slopes, drainage networks and relief distribution are described in standard texts such as those by Miller (1964), King (1966), Stewart (1968), Monkhouse and Wilkinson (1970), Pitty (1971), Young (1973) and Lewis (1974). Perhaps the most important question to ask about any such information is how does it relate to the 'ground truth'? The limitations of height data on maps are well known, but it is sometimes forgotten that in forested areas, aerial photographs show only the tops of the trees, not the underlying topography. As trees on valley floors are often taller than on slopes, the aerial photography may give a false impression of valley depth.

The fundamental data on humid landforms are the shapes and dimensions of slopes, river channels and drainage networks. The study of slope has a nodal position in geomorphology because the patterns of stream networks, the amount of dissection of relief and the steepness of slopes together characterise erosional landscapes. Most denudational processes tend to create a characteristic slope angle, which is probably the limiting angle of stability for the waste mantle of any given slope (Carson and Petley, 1970). As slope measurement techniques often seek to identify characteristic angles, the way in which slope is defined and measured is crucial (Pitty, 1966).

Equally critical in the study of humid landforms is the definition and identification of stream channels. To analyse drainage basin characteristics, the stream network must be described numerically. The system of stream ordering developed by Strahler (1952a) from that proposed by Horton (1932, 1945) defines all streams that have no tributaries as first order streams. When two first order streams join they form a second order stream; when two second order streams join they form a third order, and so on. If a first order stream enters a second order stream then there is no change in the order of the second order stream. This method of stream ordering is the most widely used in geomorphology, but alternative systems of ordering streams have been suggested by Scheidegger (1965), Woldenberg (1966) and Shreve (1967) (Fig. 7).

One problem with stream ordering is the precise definition of the first order stream. If a stream that receives no tributaries be called a first order stream, the channel of that stream must have certain basic characteristics. Although in flood times water can often be seen cascading along the floor of a grass-covered hollow above the head of any clearly visible channel, the first order stream can only be said to exist where there is a permanent clearly defined trench or trough which shows evidence of present-day use by at least occasional storm runoff. Such troughs may be only a few centimetres deep and broad and thus may not be represented on maps or visible beneath the forest canopy on air photos.

Geomorphic process surveys

Even if accurate measurements of form can be obtained, additional data on rates of change in landforms are required. Among the surveys contributing to a general geomorphological inventory are vigil network measurements. These are part of a program to record the present state of representative sections of the landscape for future generations of scientists. If precise information is avai-

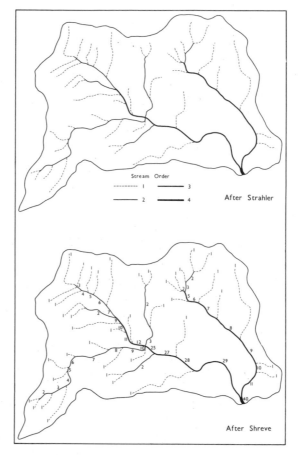

7   *Stream order in North Babinda Creek, Queensland, according to the Strahler
and the Shreve systems. North Babinda Creek has a catchment area of
15.35 km² and a perimeter of 19.31 km. From the Strahler morphometric
analyses the following indices can be derived:*

Drainage density (km/km²)	*2.46*
Relief: length ratio (km)	*0.17*
Ruggedness number (km)	*1.93*
Relative relief (cm/m)	*4.05*

*Ruggedness number is defined as the product of basin relief and drainage
density; relative relief as catchment relief divided by catchment perimeter.
While the Strahler system only creates a stream of order N if two streams
of order N–1 join, the Shreve system, which also defines each stream
with no tributaries as order 1, makes each successive link a magnitude
equal to the sum of all first order segments which ultimately feed it.*

lable on conditions and topography of a given river basin at the present time, and benchmarks are left in that area for future surveys, repeated surveys in 10, 25 or 50 more years' time will provide a record of the way in which that basin has changed with time. At present, geomorphologists have to refer to old maps, old surveys for various engineering and other purposes, and have no fixed reference points on which to base their comparisons.

Among the vigil network measurements are surveys of hillslope profiles. Resurvey along the same line will enable changes in profile to be assessed. Comparisons between slope profiles may be made by using various statistical measures of the distribution of slope angles within individual profiles.

The rate at which sediment is removed from the landscape and carried down a river system may be measured by repeated surveys of the floors of small reservoirs and farm dams, providing none of the sediment carried into the reservoir is lost by floodwater spillage over the dam. As sediment removed may be related to vegetation change, vegetation surveys form an essential part of the vigil network surveys. Of the hundreds of methods used to measure vegetation, those reporting areal coverage and weight or volume of vegetation are the most meaningful in landform studies (Emmett and Hadley, 1968). In addition to live vegetation, soil surface features such as quantities of mulch or litter, bare soil, and rock should be measured.

Descriptions of landforms require information on the materials of which the landforms are made. Some of this information may be gained from geological maps but often the scale of available geological maps is too small to provide geomorphologically significant detail. Geological maps usually indicate stratigraphic units as 'formations', which may contain a sequence of sedimentary rocks of varying lithology. As the geomorphologist requires information on lithology, he has to supplement the geological map by observations and sampling in the field and laboratory analysis of surface materials. Detailed or semi-detailed soil surveys provide excellent information on surface materials, but as most soil surveys are still at the reconnaissance stage and incomplete the landform student possessing both detailed geological and soil surveys is fortunate indeed.

The observations on materials include measurements of grain size and shape, study of petrology and chemical composition. The

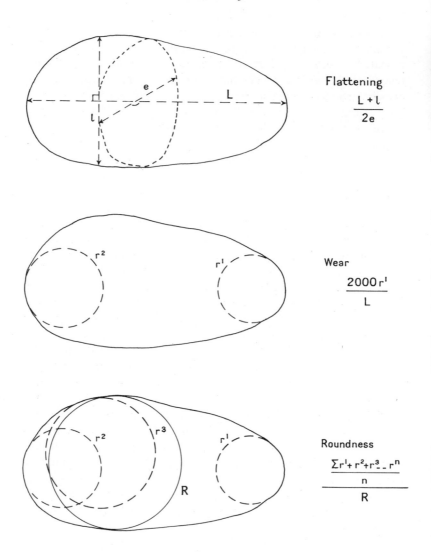

Flattening

$$\frac{L + l}{2e}$$

Wear

$$\frac{2000\,r^1}{L}$$

Roundness

$$\frac{\dfrac{\sum r^1 + r^2 + r^3 \_\_ r^n}{n}}{R}$$

8 *Pebble measurements and indices: r^1 is the radius of curvature of the sharpest corner; R is the radius of the largest inscribed circle.*

efficiency of both weathering and erosional processes depends on the size of particles released from parent rocks and exposed at the soil surface. The loads carried by rivers are in part a reflection of the size of the debris which is supplied to them.

Particle shape, described by such parameters as the index of flattening and index of roundness (Cailleux and Tricart, 1959), provides a means of characterising pebbles from different environments (Fig. 8). Although the petrographic analysis of sand grains and clay minerals involves specialised techniques that require long training and experience, the results of such analyses greatly assist geomorphological investigations.

Measurements of landform thus embrace a wide variety of observations, from simple measurements of slopes to detailed morphological mapping, from repeated surveys of marked sections to the careful sampling and measurement of sedimentary particles. Such field observations should be envisaged as part of a plan to enable the maximum possible date to be obtained from remote sensing media, particularly the aerial photograph. In suitable circumstances, even particle size and orientation measurement may be made from photographs (Caine, 1968). Processes have to be monitored and related to landforms and materials. Appropriate choice of sampling schemes and of statistical analyses of landform data is discussed in texts such as Krumbein and Graybill (1965) and Doornkamp and King (1971).

II

HYDROLOGICAL AND
BIOLOGICAL CONTROLS OF
THE PROCESSES OPERATING
ON HUMID LANDFORMS

External morphodynamic factors are the dominant cause of landform variation on features less than 10^{-2} km^2 (G-scale values of 8 and higher). Almost all the variation in humid landform processes can be explained in terms of variations in the hydrological cycle as affected by the vegetation canopy and soil mantle.

The hydrological cycle

Within a single hydrological unit, such as a drainage basin, the hydrological cycle can be viewed as a series of inputs (precipitation) which take various routes through the basin to outputs as channel runoff, evapotranspiration and groundwater discharge. The rain may be intercepted by the forest canopy, whence it may evaporate back to the atmosphere, fall as canopy drip from leaf tips to the forest floor, or flow down the trunk of the tree as stem flow (Fig. 9). The tree canopy can only retain a finite amount of water, and as the intensity and duration of the storm increases more and more of the rain merely collides with the canopy and then falls to the ground together with the stemflow and the raindrops which pass through spaces in the canopy as throughfall.

In deciduous forests, there is a great seasonal contrast in interception, and consequently in the routes taken by precipitation inputs, between the periods when the foliage has its fullest development and the winter months when the trees are bare of leaves. In tropical rain forests, and in most temperate evergreen forests, where the density of the foliage and the form of precipitation remain the same throughout the year, the proportion of the rain intercepted varies only as a function of the duration and intensity of the rain.

On reaching the ground, rainwater either percolates down into the soil or runs over the surface as overland flow, some of which may infiltrate further down slope, while the remainder reaches a permanent river channel. As the forest soil is broken up by humus and bacterial action, and has many root channels and animal

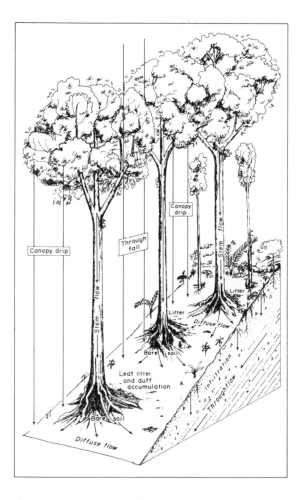

9 *Flow of water through the vegetation canopy and the soil in a tropical rain
   forest (drawing by A. J. King)*

burrows, a large proportion of the water reaching the ground in
the forest infiltrates.

Some of the water which infiltrates will continue down to the
unweathered rock, where, depending on the nature of the rock, it
may either find joints and cracks through which it may penetrate,
or pores in the rock itself through which it can seep under
the influence of gravity until it reaches the zone where all the pore

spaces in the rock are fully saturated with water. The upper surface of this zone of saturation is termed the water table. As the groundwater in the zone of saturation moves under the force of gravity very slowly through the rock, the water table actually maintains a sloping surface, highest beneath the hill tops and divides, lowest in the valleys. Depending on the geological structures beneath the drainage basin, deep groundwater circulations may transfer water across the watershed divide, with flows moving both up and down under hydrostatic pressure along hydraulic gradients.

Not all the water penetrating the surface of the soil reaches the water table, some may be taken up by plants and transpired, while other water may move laterally down slope within the soil as throughflow (Fig. 9) or may be held, after most of the water has drained away, by capillary tension, tiny films of water adhering to grains, particularly at points of grain contacts, where they remain until absorbed by plant rootlets or lost by evaporation.

Not only the large trees but all plants and animals of the forest community are involved in the consumption, temporary storage, and disposal of water. This biotic activity may be illustrated by three typical forest ecosystems: tropical rain forest, temperate deciduous forest, and cool temperate coniferous forest.

Tropical rain forest

Rain forest is remarkable for the complexity of its plant community, often with over 400 different tree species occurring within a square kilometre, and for the density of the vegetation cover. The tallest trees in a lowland equatorial rain forest (Fig. 10a) may reach almost 50 m above ground level, forming a continuous canopy which greatly reduces the amount of light reaching the lower layers of the forest. Sometimes, as in lowland dipterocarp forest in Malaysia, emergent trees stand out above the general level of the canopy. Trees which fail to reach the general closed canopy height are many and include both young emergent trees growing up towards the light and other species adjusted to a lower light need. Festooned on some trees are creepers (vines) and epiphytes, all of which add to the variety of plants competing for light and water. On the forest floor are the young shoots of new trees, occasional masses of tangled vines where the larger trees have fallen and the light has reached the forest floor and caused rapid growth of vines, fallen logs and decaying wood and leaves (Plate 1). The rate of decay, aided by

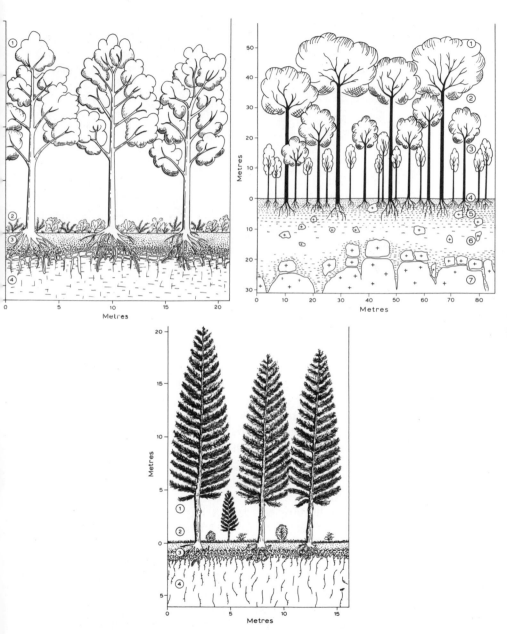

10   *Schematic forest profile diagrams. See pp. 22–5.*

1　*Litter-covered floor of rain forest on Mt Duau volcano, Papua New Guinea.*
Soil and litter collects on upper sides of roots as a result of diffuse slope
wash (CSIRO photo).

termites, bacteria and fungi, is rapid, so that the organic material
on the ground surface is thin and easily washed away to expose the
soil underneath. The tree roots spread laterally, but may reach up
to 10 m down into the weathered rock material, though the depth
of the weathered rock may reach 30 m.

Biologically, the humid tropical rain forest is two to three times
as productive as cold temperate forest, the average total green
matter production at two sites in Ghana and the Congo being 27·9
metric tons per hectare per year (t ha^{-1} y^{-1}) compared with an
average of 11·7 t ha^{-1} y^{-1} in cool temperate forests (Bray and
Gorham, 1964). But while the humid tropics support the most
productive forest communities of the world, the productive effort
is largely consumed in self-regeneration, most of the nutrients
falling to the forest floor being re-used by the growing plants in the
forest. Thus the geochemical cycle in the forest parallels the
hydrological cycle in that much of both nutrient matter and water
falling to the forest floor is taken up and used in the growth of plants.

Temperate deciduous forest

This forest type, typical of eastern North America, western Europe and eastern Asia, is strongly contrasted with both temperate and tropical evergreen rain forest in that it consists largely of single species dominant stands. In Britain, the oakwoods (*Quercus robur*) are typical of the clayey brown forest soils, with beech (*Fagus*) on the lighter calcareous soils, particularly on the chalk lands. Here the single dominant trees form one canopy layer (Fig. 10b), with a dense and diverse ground cover of shrubs and herbaceous plants. With the seasonal rhythm of the temperate climate, leaf fall in the autumn supplies large quantities of organic matter to the ground surface, which does not decompose rapidly in the cold winter months, and through which emerge early flowering spring plants during the first warm months when daylight begins to exceed 12 hours. As the leaf cover develops, the spring flowering plants die and add to the accumulation of organic matter which decays through the activity of organisms and biochemical processes during the summer. The weathered rock and soil layer is much thinner than under the tropical forest, and tree roots often penetrate deep into cracks in the bedrock.

Cool temperate coniferous forest

Coniferous forest consists of dense stands of straight-trunked conical trees with relatively short branches and small, narrow needle-like leaves. Where evergreen, the coniferous forest provides continuous and deep shade to the ground so that lower layers of vegetation are sparse or absent, except for a thick carpet of mosses in places (Fig. 10c). Like the deciduous forest, large areas are covered with trees of the same species. Long, cold winters and relatively cool summers preclude effective bacterial action and thus thick accumulations of organic matter build up, particularly where drainage is poor. The trees are shallow rooted, and the depth of weathered rock material beneath them is slight, so that occasional rare but strong winds can cut broad swathes of trees felled by wind blowing through the forest.

ORGANIC MATTER IN GEOCHEMICAL PROCESSES

Living organisms represent a small mass compared with the rock mass of the earth, but in the vital surface layer of the earth where landforms evolve, living matter is being created and destroyed continuously. In humid climates, this biological cycle of activity is the dominant control of the geochemical processes, including chemical weathering and soil formation, operating in the outer layer of the earth's crust.

The formation of living matter from organic compounds existing in the environment is largely the result of the metabolism of green plants. These synthesise organic compounds from CO_2 of the air, water, and mineral salts. In photosynthesis carbohydrate is formed from six molecules of CO_2 and six molecules of water, using up 674 calories of energy. These materials are combined by the action of chlorophyll, the green pigment in plant leaves. New, complex compounds are formed in this way. Calcium, magnesium, potassium, iron, and other elements are sorbed from soils to manufacture complex organometallics. Thus complex biogenic accumulation of chemical elements occurs in organic matter. The uptake of nutrients by vegetation from the soil is one part of the process of transformation and translocation of rock material involved in landform development.

Animals, certain plants such as fungi, and most micro-organisms are not capable of organic synthesis from inorganic matter. However, some micro-organisms synthesise their organic compounds from mineral substances, not with sunlight, but with energy liberated during some chemical reaction, e.g. the oxidation of iron.

Although all chemical elements pass through the stage of incorporation within living matter, this is more typical of some elements than others. Nitrogen, potassium, phosphorus and even sulphur are constantly tied up in organic complexes. Oxygen is the most abundant element in all organisms and is followed in descending order by C, H, Ca, K, N, Si, P, Mg, S, Na, Cl, and Fe.

Plants are thus storing elements they have taken up from the soil and are making them available to animals in an assimilable form. Animals may take matter from one place to another, or in special circumstances, such as the action of termites or corals, accumulate it in specific localities. Destruction of organic matter by decomposition, particularly through the action of micro-organisms, results in the destruction of proteins, fats, cellulose and other organic substances and in the production of carbon dioxide and organic acids, important in weathering.

Destruction of organic matter is not the only function of micro-organisms, which operate at all stages of the biological cycle. Microscopic fungi which live on the surface of rocks attack and dissolve part of the minerals which make up the rock structure (Silverman and Munoz, 1970). Citric acid is secreted over the rock surface by the organism, and the acid attacks metallic salts — especially those of iron and magnesium — in the rock minerals, breaking down the mineral's crystal structure and making the salts soluble in water.

While plants need sunlight for photosynthesis and living matter can survive only at the earth's surface or in cavities below ground where oxygen can be obtained, decomposition of living matter can take place at considerable depths below the surface. Microbes have even been found in oilfield waters from several thousand metres below the surface. Thus changes in rock formations are taking place as a result of organic activity not only at the rock/atmosphere inter-face but also at considerable depths below the surface.

Seasonal variations in biotic activity

Even in the equatorial rain forests seasonal rhythms occur. Periodicity in tree flowering has been noticed many times in Malaysia, where a drier period than usual seems to stimulate flowering, but such stimuli do not occur regularly every year. Within the tropics, the seasonal shifts in the atmospheric circulation can cause storm tracks to move away from zones where they are usually persistent and for undisturbed air streams to be established for periods of one month or more. Thus at many tropical stations, where annual rainfall varies little, the length of dry periods varies considerably from one year to another. This variation in precipi-tation can be reflected by periods of biological activity, either production of new green matter in wet years, or, possibly, flowering in dry years.

Once the equatorial regions are left behind, seasonal rhythms become more and more marked. At 17°S in the rain forests of coastal Queensland near Innisfail, although there is always enough water to support the rain forest vegetation, a marked contrast occurs between the summer wet season, when the NE winds affect the area, and the winter season when the SE winds bring less rain. Rain forest is found at intervals all the way down the east coast of Australia into Victoria and then again on the west coast of

Tasmania. On the southern island, the thermal contrasts between summer and winter are more important controls on the rhythm of biological activity than precipitation contrasts. Jackson (1965) suggests that in Tasmania the limiting factor for rain forest is 50 mm per month of summer rainfall, rather than an annual rainfall total exceeding 1400 mm.

The seasonal contrasts in temperature and significance of rain and snow in cool temperate deciduous and coniferous forests are associated with short winter and long summer hours of daylight, the summer being a period of intense biotic activity, with rapid vegetation growth. Many animals are as adjusted to the seasonal rhythm as the plants, hibernation being commonplace among the smaller mammals of the deciduous and coniferous forest. The materials of the ground surface undergo two kinds of processes, in summer dominantly chemical and biological activity, restricted only by occasional shortage of water due to the high demand for evapotranspiration, and winter processes affected by snow, frost, melt water and high soil moisture levels. These contrasts in hydrologic and biotic activity in the tropical and temperate forests exert an important influence on the weathering and erosion processes described in subsequent chapters.

III

WEATHERING

Weathering is the sequence of changes affecting materials near the earth's surface during adjustment to environmental conditions prevailing at the atmosphere-lithosphere interface. It occurs when rock materials become exposed and subject to surface temperatures and pressures. Minerals and rock materials which were stable at higher temperatures and pressures deep in the crust may be relatively unstable at the earth's surface and are thus subjected to a series of physical and chemical reactions which rarely operate alone. These basic processes are often illustrated from relatively rare situations where a single process is dominant and not from the more common situations where weathering is the result of complex interacting processes.

PHYSICAL WEATHERING

At some localities within forested areas large masses of bare rock project out of the forest canopy or may be exposed by hurricane destruction of the trees. On these bare rock surfaces the process of physical weathering, which breaks up the rock without destroying its chemical composition and without the action of water on the rock materials, may operate unaided by other types of weathering. These processes include expansion resulting from unloading, crystal growth, thermal expansion, organic activity and colloid plucking.

Igneous rocks formed beneath the surface at conditions of high temperature and pressure are in a slightly compressed state when deeply buried. When landforming processes remove the overburden, these rocks adjust to their exposure by a slight expansion in volume which results in the breaking off of sheets of rock parallel to the ground surface. In some circumstances this gives rise to bare domes of rock, where sheeting has produced rounded exfoliation landforms. However, while this unloading or sheeting may create landforms in dry climates on its own, chemical processes have helped to form bare domes in humid areas.

29

Unloading and sheeting may help to create fissures and cracks into which water can penetrate to further the destruction of the rock. In winter in cool temperate climates such water may freeze and break up the rock by frost wedging. Frost and ice also affect the movement of soils and debris on slopes by periglacial creep, frost sorting and allied processes (Davies, 1969).

If a saline solution enters pores in the surface of a rock, and several days of strong evaporation without rainfall or spray occur, crystallisation of salt may take place and promote granular disintegration of the rock surface (Evans, 1970). Although most likely to occur in dry areas, salt crystallisation can occur in humid regions on exposed rock surfaces that dry rapidly. The surface of a rock exposed to the sun may be affected in this way, while the damper, shaded part of the rock undergoes chemical disintegration. Such local and microclimatic contrasts are of great significance in weathering processes generally.

The mechanical disintegration of rock owing to the expansion and contraction of alternately heated and cooled surfaces is thought to be important by Ollier (1963, 1965, 1969a). The stresses set up by the heating of the outer layers or the exposed upper parts of rocks can produce deep cracks which eventually split the boulders apart. However, Blackwelder (1925, 1933) showed that the presence of water was important for the cracking of rocks under alternating hot and cold thermal conditions, while Chapman and Greenfield (1949) found that spheroidal scaling involved the oxidation and hydration of silicate minerals. Although bare rock surfaces in humid regions experience the wide diurnal temperature ranges typical of desert areas, they are wetted and dried far more frequently than desert rocks, and thus rock disintegration is more likely to involve chemical changes as well as physical weathering.

Organic activity breaks up rocks by the widening of fissures by growing plant roots and through the burrowing activities of animals and insects. Termites in the Brocks Creeks region of the Northern Territory move about 0.48 m^3 of earth per ha annually (Williams, 1968). It seems probable that soil colloids may have the power to loosen or pull off small bits of rock from the surfaces with which they come into contact (Thornbury, 1954), but as such colloidal plucking may involve complex biochemical reactions it may not be purely physical weathering.

CHEMICAL WEATHERING

Continually changing weather and hydrological conditions at the earth's surface force surface mineral materials to seek new equilibria through chemical weathering processes. The water that reaches the ground surface is not pure, containing varying amounts of natural and man-made contaminants. Contact with natural outcrops of sulphate and sulphide minerals renders surface waters acid and generally throughout north-western Europe sulphur compounds dissolved from the atmosphere cause rainwater to act as a dilute concentration of sulphuric acid affecting the weathering of limestones (Sweeting, 1966) as well as the decay of building stone as at Strasbourg Cathedral (Millot *et al.*, 1967).

In water and aqueous solutions a few water molecules are split into H^+ and OH^- ions. The product $H^+ \times OH^-$ is always 10^{-14}. In neutral pure water there are equal numbers of H^+ and OH^- ions and the pH, the negative logarithm of the H^+ concentration, is 7. Acid solutions have more H^+ than OH^- ions and thus pH less than 7; alkaline solutions have more OH^- ions and thus pH greater than 7. Rainwater generally has a pH slightly below 7 (Table 3), low pH values in British samples arising from the sulphur pollution just described. Rainfall acidity, a major factor in rock weathering, is increasing as a result of human activity, as demonstrated by Likens and Bormann (1974) in North America.

TABLE 3 **pH of precipitation at tropical and temperate localities**
(after Douglas, 1968b)

Tropical localities	pH range	Temperate localities	pH range
Singapore	4.7–8.2	Plymouth (U.K.) (mean)	4.0
Sarawak	4.9–6.0	Lake District (U.K.)	4.0–5.8
Selangor	5.1–7.2	Ingleborough (U.K.)	4.0–7.1
Ivory Coast	5.1–7.4	Scandinavia	4.8–6.6
North-east Queensland	5.2–6.6	North Humberside (U.K.)	5.1–6.7
Surinam	5.3–6.9	Kentucky (U.S.A.)	5.5–6.2
El Salvador (mean)	5.5	Australian Capital Territory	5.6–6.7
		Downderry (U.K.) (mean)	6.6

The solubility of many substances is affected by pH. Ferrous iron is much more soluble at pH 6 than at pH 8·5. Silica on the other

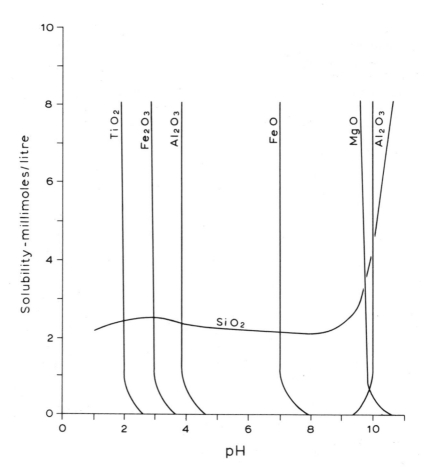

11 *Solubility of various rock components as a function of pH*

hand increases in solubility rapidly above pH 9, but below pH 9 factors other than pH many affect silica solubility (Siever, 1959) (Fig. 11).

Solution is commonly a first stage of chemical weathering. Data are available for the solubility of various elements in water, but as the water involved in the weathering process is never pure, but usually a weak acid solution, the solubilities of materials are higher than these data indicate. Solution depends on the rate at which water reaches the solid/liquid interface. Conditions that allow rapid water movement favour rapid solution. More solution may thus take place at the surface of the soil, where loose fragments are quickly wetted on all sides by rainwater, than in the cracks and fissures of the underlying rock through which water moves but slowly. However, solution may be greater on rock fragments within the soil beneath vegetation.

The stability of elements in any particular oxidation state (for example iron in the ferrous or ferric state) depends on the energy change involved in adding or removing electrons. This is measured as the oxidation potential and is expressed in terms of Eh. The Eh varies with the concentration of the reacting substances, and if H^+ or OH^- is involved, then the Eh varies with the pH of the solution. In aqueous conditions, the Eh becomes lower as pH increases, and so oxidation proceeds more readily the more alkaline the solution. Thus in the acid environments associated with some types of swamp such as the lowland peat swamps of the Malay Peninsula and of Borneo, oxidation is not occurring, while in highly alkaline environments oxidation is extensive (Fig. 12).

Oxidation processes take place in the zones of wetting and aeration. Iron oxides, prominent features of profiles in all humid regions, frequently occur in a band about 100 mm thick on exposed rock surfaces in the humid tropics while below this layer the rock retains its original colour. In bauxites, aluminium occurs in hydroxides such as gibbsite whose structure consists of layers composed of two close-packed hydroxyl sheets with aluminium atoms lying in six-fold co-ordination between them.

Many compounds containing water are not definite chemical entities, but have the molecules of water attached very loosely to the structure. Such compounds are termed hydrates, the hydration being the chemical combination of water and another substance, often producing an increase in size which when combined with thermal expansion and contraction effects may cause rupture along fracture planes, sheeting parallel to the surface and granular disintegration. Hydration should not be confused with hydrolysis. During hydration there is no selective consumption of H^+, and cations are not released to the altering solutions. The reaction of aluminium oxide and water to form gibbsite is hydration

$$Al_2O_3 + 3H_2O = 2AL(OH)_3$$

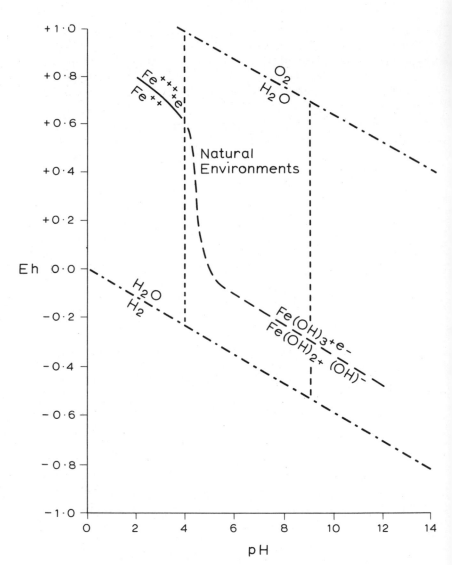

12 *Eh-pH diagrams (after Garrels and Mackenzie, 1971):*
    (*a*) *Eh-pH diagram for ferrous and ferric hydroxides. The solubility of
    ferrous iron is in part determined by the reducing power of the environ-
    ment. Thus the change in stability of ferric (Fe⁺⁺⁺) and ferrous
    (Fe⁺⁺) iron over the pH range 4 to 6 changes rapidly as Eh conditions
    alter. Ferric compounds are stable in the environments above the line,
    ferrous combinations in those below.*

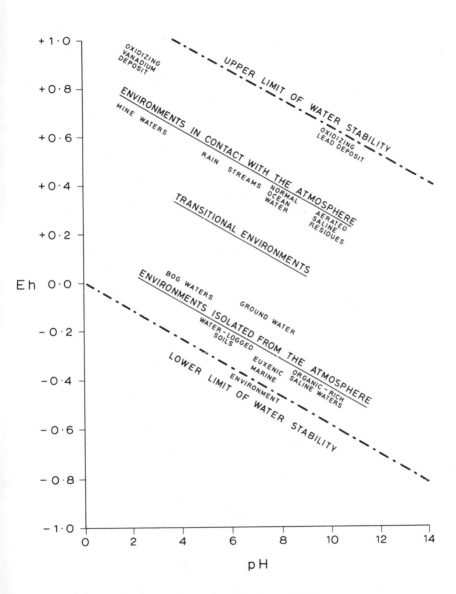

12 *Eh-pH diagrams (after Garrels and Mackenzie, 1971):*
   (b) Eh-pH diagram indicating alkalinity, acidity, oxidising and reducing
   conditions associated with various environments.

Most silicates and oxides can be decomposed, and compounds such as clays produced, by hydrolysis in which the H^+ ion enters the atomic structure of a mineral in exchange for a cation. However, to decompose a mineral completely, the hydrolysis reaction must be coupled with the interchange reaction whereby the absorbed H^+ on the surface of a crystal diffuses to the interior; in turn an interior cation diffuses outward to occupy the position of the cation which was replaced by the H^+. Hydrolysis and the interchange reaction can occur until the mineral is completely decomposed (Barshad, 1972). Two types of hydrolysis reaction are involved in weathering a feldspar such as orthoclase. In the first a soluble base is separated from an insoluble fraction: $K(Al\ Si_3O_8) + H^+ + OH^- \longrightarrow H (Al\ Si_3O_8) + K^+ + OH^-$. The interchange reaction follows: $H(Al\ Si_3O_8) \rightarrow Al\frac{1}{3}\ (HAl\frac{2}{3}\ Si_3O_8)$. Then the second hydrolysis reaction creates silicic acid and aluminium hydroxide (gibbsite): $Al\frac{1}{3}\ (HAl\frac{2}{3}\ Si_3O_8) + 3H^+ + 3OH^- = Al\ (OH)_3 + (H_4Si_3O_8)$.

Igneous rocks contain silicate minerals which formed from high temperature liquids under much greater pressures than those at which they are weathered. Basically, igneous minerals consist of silicate tetrahedra in which one silicon atom is at the centre of four oxygen atoms (Fig. 13). The silicon atom of valence 4 is sometimes replaced by an aluminium atom of valence 3, but of similar size, thus not causing serious disruption of the lattice. However, the lower valence means that other atoms such as potassium or calcium must enter to maintain the electrical balance of the lattice, as in the feldspars, which are composed of such a framework with aluminium replacing silicon and alkali or alkaline earth cations entering the lattice to balance the deficiency of positive charge. Albite, sodic plagioclase, $Na\ (Al\ Si_3O_8)$, is an example of an Na^+ Al^{3+} substitution for a Si^{4+} ion (Searle and Grimshaw, 1959). The resistance to weathering of silicate minerals depends greatly on the number and kind of such substitutions.

The nature of clay minerals produced by the weathering of these minerals depends on the parent mineral and the environmental conditions. As clay mineral grains are only of the order of 1 nanometer (10^{-9} m) in diameter, they can only be photographed with the aid of the electron microscope. They consist of flat, platy layers of alumino-silicates. Being small, yet having a large surface area in relation to mass by virtue of their shape, clay minerals are extremely sensitive to conditions of intense chemical weathering and are

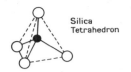

Silica
Tetrahedron

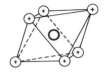

Alumina
Octahedral unit

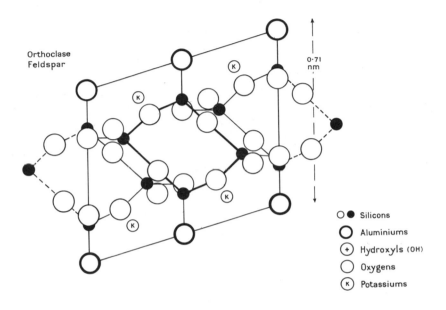

Orthoclase
Feldspar

0·71
nm

○ ● Silicons
◯ Aluminiums
⊕ Hydroxyls (OH)
○ Oxygens
Ⓚ Potassiums

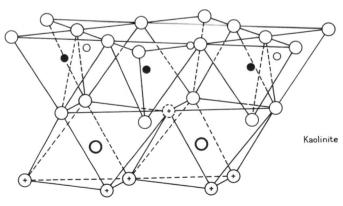

Kaolinite

13 *A comparison of the structure of orthoclase feldspar and kaolinite to show
the contrast in aluminium bonding*

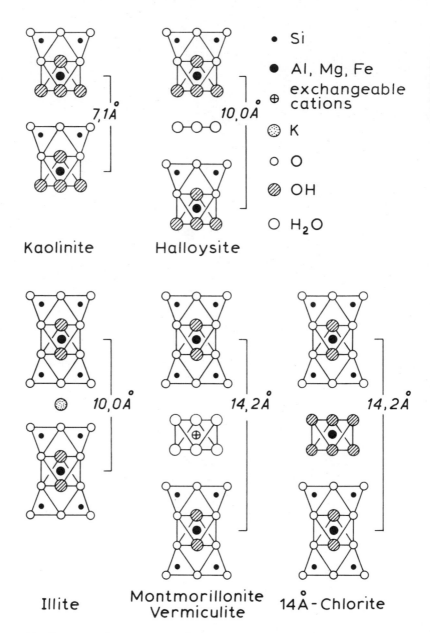

14 *Structure of some common clay minerals. Sizes are indicated in Angstrom*
    units (1A = 0.1 nonometres).

therefore a valuable indicator of conditions of erosion or sedimentation.
The stability of the silicate minerals to weathering is shown below:

Increasing resistance ↓	Olivine	Calcic plagioclase
	Augite	Calcic-alkalic plagioclase (Anorthite)
	Hornblende	Alkali-calcic plagioclase
	Biotite	Alkalic plagioclase (Albite)
	Potash feldspar	
	Muscovite	
	Quartz	↓

Clay minerals are phyllosilicates with a sheet structure somewhat like that of micas. They consist essentially of two types of layers. One is a silica tetrahedral layer consisting of SiO_4 groups linked together to form a hexagonal network of the composition Si_4O_{10} repeated indefinitely. The other type of layer is the alumina or aluminium hydroxide unit, which consists of two sheets of close-packed oxygens or hydroxyls between which octahedrally co-ordinated aluminium atoms are embedded in such a position that they are equidistant from six oxygens or hydroxyls (Fig. 14).

Clay minerals may be divided into two layered and three layered alumino-silicate sheet structures. The simpler kaolinite group have an aluminium layer and a silicate layer (Fig. 14). The other group consists of two layers of silicates sandwiching an aluminium layer together in various ways (Fig. 14). In the mineral montmorillonite, they are held together by water layers, while illite has potassium linking the layers. Chlorite consists of the basic montmorillonite structure with $Mg(OH)_2$ sheets between the three layer units.

Because of the common basic alumino-silicate structure of clay minerals, one clay mineral may evolve into a different mineral if transferred to a new site or subjected to environmental change. The expansion of the lattice may allow different bonding ions to come between the aluminium silicate layers, or for such ions to be lost (Fig. 13). The relative resistance, increasing from stage 1 to stage 13, to weathering of clay-sized minerals is set out below, with the most common clay minerals in italics.

Weathering stage	Mineral
1	Gypsum (also halite, etc.)
2	Calcite (dolomite, aragonite)
3	Olivine-hornblende (diopsite)
4	Biotite (glauconite, *chlorite*, antigorite)
5	Albite (anorthite, microcline, stilbite)
6	Quartz (cristoballite)
7	*Illite* (muscovite, sericite)
8	Hydrous mica intermediates
9	*Montmorillonite* (beidellite)
10	*Kaolinite* (halloysite)
11	Gibbsite (boehmite)
12	Haematite (goethite, limonite)
13	Anatase (rutile, ilmenite, corundum)

The aluminium and iron oxides, which are more resistant than kaolinite, have a simpler structure, and thus it is possible to think of the evolution of clay minerals as the gradual simplification of structure as the weathering environment becomes increasingly severe. However, it is equally important to note that a variety of combinations of clay minerals occur in nature. Interstratified illite-chlorite minerals or montmorillonite-kaolinite minerals may be found (Tricart, 1965a). Various ions may also become lodged in a thin wedge between layers of a clay mineral, for example Ca^{++} and Na^+ ions. These modify the properties of the clay mineral, particularly the mechanical properties. Ca^{++} ions increase the cohesion of the mineral, while sodium ions decrease it, calcium helping to aggregate the soil, and sodium causing swelling under wet conditions.

The influence of past conditions on the type of clay minerals found in a particular locality may be strong. Thus the factor of inheritance is important in considering the weathering profiles found on sedimentary rocks, particularly in temperate latitudes where climates have been unstable in the recent geological past. Lucas (1962) and Millot (1965) distinguish between the degradation of clay minerals in leaching environments during pedogenetic processes and the aggradation of clay minerals after deposition in ill-drained or wet environments (Fig. 15). The degradation of clay minerals occurs when the basic structural framework of silicate mineral remains more or less intact, but with considerable alteration.

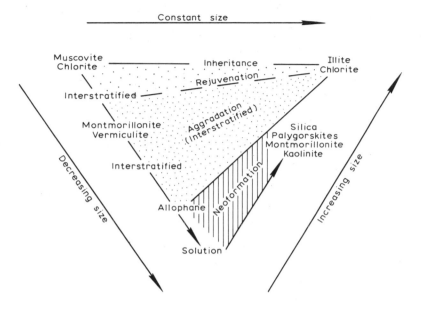

15 *Patterns of evolution of clay minerals (after Millot, 1965).*

The aggradation of clay minerals occurs in depositional environments, particularly where clays brought down by rivers are transformed in the brackish water conditions of deltaic areas. Here riverborne montmorillonites and illites, usually deficient in potassium and magnesium, are aggraded by the uptake of these cations in the nearshore zone. Their lattices are thus restored and the minerals may even regain the composition of chlorite (Porrenga, 1967). Such processes may be visualised as a sub-cycle of the geochemical cycle as indicated in Figure 6.

The neoformation of kaolinite is characteristic of the weathering of granitic rocks. On the acidic rocks of Emma Range in Surinam, direct kaolinite formation does not always take place, the feldspars being degraded immediately into gibbsite by intensive weathering under conditions of high pH and slight percolation (Wensink, 1968). The pH on the surfaces of mineral grains becomes high enough for aluminium to go into solution, but the aluminium is redeposited as gibbsite in the immediate vicinity of the crystals. Some of this gibbsite may be resilicified to form kaolinite.

On the local scale, where lithology and drainage conditions are more or less uniform, differences in precipitation produce considerable contrasts in soil mineralogy. The kaolinite concentration in the krasnozemic basaltic soils of the Atherton Tableland in north Queensland varies from 70 to 80 per cent in the areas with the lowest mean annual rainfalls (900 mm) to about 50 per cent in the wettest places (3500 mm) (Simonett, 1960). A parallel situation occurs in Hawaii where at moderate rainfalls (mean annual less than 800 mm) montmorillonite is the principal mineral. An increase in annual rainfall results in kaolinite formation, while in the wettest areas aluminium, iron and titanium oxides are formed and little montmorillonite or kaolinite is found (Sherman, 1952; Degens, 1965). This succession of clay-sized mineral formation corresponds to the stages of weathering of such minerals discussed earlier.

Where lithologies are variable, and where drainage conditions differ, wide local contrasts in the clay minerals in weathering profiles are to be found. The significance of drainage in weathering processes may be illustrated by the bauxite deposits at Mungu Belian in Sarawak. Here a 3 m thick bed of bauxite with a high alumina content has formed under a thin soil cover on hills which rise about 25 m above the surrounding alluvium. Beneath the bauxite is massive, soft, buff kaolinitic clay, in places over 20 m thick. Although silica is virtually absent from the bauxite, it is a major component of the kaolinitic clay and makes up about 52 per cent of the parent andesite. The bauxite developed on well drained sites above the water table where groundwater could circulate freely and dissolve large amounts of silica, but below the water table, where water percolating through the rock contains considerable concentrations of dissolved salts, kaolinitic clay formed from the same parent material.

Thus variations in lithology, drainage, and topographic situation cause differing patterns of weathering under the same climate. Climate itself affects weathering, with more clays being formed in higher temperature and wetter localities. This diversity of clay mineralogy has important effects on the behaviour of the weathered mantle covering fresh rocks in relation to the processes of erosion. Clay minerals are thus an important aspect of the physical and chemical properties of weathering profiles, as well as being useful clues in the elucidation of the history of landforms.

Weathering of intrusive rocks

Broadly similar decomposition products are evolved on granite in all humid climates. Under humid tropical conditions granite weathers into a reddish or yellowish brown clayey sand or sandy clay virtually free of particles of unweathered rock, above which may be a migratory layer of colluvium capped by organic soil. In the sedentary profile, the weathered material retains the structure of the original rock, yet is quite soft, yielding readily to the impact of a spade. Feldspars and biotites are transformed into montmorillonites and, more commonly, kaolinites, while the quartz grains remain hard and unaltered. A bi-modal grain size distribution of the weathered material results, a granite-derived soil profile near the Gap Road, Bukit Fraser, Malaysia, for example, having 31 per cent clay, 21 per cent silt, 14 per cent fine sand, 20 per cent coarse sand and 14 per cent gravel (Roe, 1953). The weathered debris from granite also contains large boulders of unweathered rock, known as corestones. These corestones range from less than 1 to over 10 m in diameter and are more frequent towards the base of the profile. In some places, the transition from weathered to unweathered granite is abrupt, and a basal surface of weathering can be recognised easily. Elsewhere the transition from boulders to fragments of rock separated from the main body of unweathered rock only by decomposition along joints is gradual.

In places where the grit, the sand and clay-sized weathered material has been removed, slopes mantled by corestones occur. In coastal Queensland such slopes may be the result of destruction of the forest by tropical cyclones and exposure of the grit to slopewash. In many valley heads in rugged granitic terrain, the slopes at the head of the stream are a mass of corestones left behind as the intervening grit has been removed by slopewash, throughflow and stream erosion.

Bare granitic domes occur occasionally in humid environments. Chemical weathering on such surfaces produces *Silikatkarren* (Tschang Hsi-Lin, 1962; Jennings, 1971), rounded grooves in the rock surface (Plate 2), probably caused by solution of silica under high pH conditions by water whose alkali concentration has been increased by rapid evaporation on the hot, bare rock surfaces. On such bare surfaces, too, there develop thicker weathering rinds than occur on corestones buried in the weathering mantle. Such

2   Silikatkarren, *Trengganu, Malaysia (Photo by J. N. Jennings).*

weathering rinds are analogous to the hardened outer layers developed on exposed rock surfaces in hot arid climates, for the microclimate of the bare rock surface is subject to the same intense heating
and drying as desert rocks.

Weathering of volcanic rocks

When magma is ejected through volcanic vents on to the surface
of the earth, it cools more rapidly than the intrusive magmas from
which coarse grained granites are formed. The fine grained rocks
that result are usually of andesitic or basaltic type, poor in quartz
and with moderate to large amounts of ferromagnesian minerals,
the basic basaltic rocks containing the higher proportions of the
latter minerals. The basic rocks weather more readily than most
other igneous rocks. Both ferromagnesian minerals and feldspars
are easily altered to clay minerals, usually kaolinite in warm, humid
climates, but often with the development of iron oxides or
aluminium oxides which produce lateritic or bauxitic concretions
in the characteristic deep weathering profiles. In humid climates
recolonisation of volcanic slopes by vegetation begins almost as
soon as the ash is cold. Despite this rapid rate of weathering, on
steep volcanic slopes on Bougainville and Buka Islands, Papua New

Guinea, the removal of surface material is equally rapid and deep weathering profiles do not develop (Speight, 1967a). Where the rocks are older, as in the Kuantan district of West Malaysia or the Atherton Tableland of north Queensland, uniform deep red-brown weathering profiles develop, with much less contrast in texture than in granite-derived weathering products. Andesitic soils tend to have a bimodal grain size distribution, the silt and clay fractions dominating, while basaltic soils are dominated by clay size particles.

In temperate environments, the weathering of these volcanic rocks may not proceed to the extent just described, the formation of montmorillonitic clays being common. Montmorillonite derived from basalt weathering gives rise to some of the 'cracking clays' of the Merriwa plateau area of the Hunter Valley, New South Wales, where mean annual precipitation is only 550 mm. In the same region, under wetter conditions, on the margins of the Barrington Tops, where the mean annual precipitation exceeds 1000 mm, kaolinitic clays are formed (Van de Graaf, 1963).

Weathering of sedimentary rocks

The materials of which clastic sedimentary rocks are composed have undergone phases of weathering, erosion and transportation before being included in the formations in which they are now found. Thus minerals that are resistant to weathering make up most clastic sedimentary rocks, quartz being the chief constituent of most sandstones and an important constituent of siltstones and shales. Feldspar, although the most abundant mineral of igneous rocks, plays a role subordinate to that of quartz in the sediments.

In sandstones, weathering consists primarily of the destruction of the cement holding the quartz together. If jointed, the sandstones will be broken up primarily along the joints, giving slabs of rock which weather slowly, as occurs on the Hawkesbury Sandstone around Sydney. In humid conditions, where chemical weathering is dominant, sandstones form relatively resistant rocks unless they have soluble cements, such as the limestone cement of calcareous sandstones. Shales on the other hand consist largely of laminae of clay minerals. In the humid tropics the claystones, mudstones and schists are transformed from hard rock to a soft clay mass, with mudstones retaining their laminated appearance and schists their foliation. Immediately above the partly weathered parent material may be fragments of ironstone or laterite having the same layered structure. These fragments are thought to be formed in the incompletely weathered zone by displacement of the rock minerals by iron, retaining at first the original foliated structure of the rocks and gradually increasing in thickness by further deposition of iron to form the roughly flattened ironstone fragments usually associated with soil derived from schists (Owen, 1951).

Quartz, derived from hydrothermal mineral veins, and quartzite, metamorphosed sandstone, are almost chemically inert and weather primarily through physical processes. Quartz and quartzite thus form upstanding resistant landforms in the humid tropics, such as the Klang Gates Ridge near Kuala Lumpur, West Malaysia. In such countries outcrops of quartz provide rare patches of whitish-grey bare rock projecting through the general green of the jungle canopy. While much smaller in size, quartz veins through metamorphosed rocks form upstanding ridges on exposed rock surfaces in humid temperate environments, as exposed for example below Torrs Walk at Ilfracombe, England.

Quartzitic sandstones are common in many areas of western Europe and around Sydney, New South Wales. Where still horizontally displaced, such sandstones form the horizontal uplands of the tablelands, creating the spectacular Blue Mountains at Katoomba and the escarpments inland from Wollongong. These sandstone features recur in the Pennines, North York Moors, Ardennes, Vosges and Bohemian Massif of western Europe. The edges of the sandstone plateaux of these areas are marked by outcrops of broken, weathered rock, sometimes with residual pillars (tors) isolated from the cliff face itself (Plate 3). Similar landforms occur in a variety of climates (Mainguet, 1972) but in temperate latitudes they are probably inherited from past climates with intense chemical weathering and modified by physical processes, particularly frost action in the winter at present and even more so in cold periods during the Quaternary. The detailed pitting of sandstone and conglomerate landforms is primarily the result of differential weathering of the minerals contained in the rocks. Where the composition of sandstones varies, the strata with more soluble cement are preferentially weathered, giving rise to slabs of resistant rock separated by almost completely weathered bands as in the sand rock cliffs of the Ashdown Forest of southern England (Robinson, 1971).

3　*Three Sisters at Katoomba, New South Wales*

Soil formation and the regolith

The breakdown and transformation of rocks and rock minerals is a preliminary to the development of soils. However, the soil mantle contains other matter and is defined as:

the natural weathered material in which plants grow and by which they are supported and supplied with both water and mineral foods. The constituents of the soil-mantle are mineral particles—clay, silt and sand—incorporated with air, water and humus, the last-named providing a medium for bacterial activity and, in some cases, forming a recognizable horizon, or more than one (Brade-Birks, 1946: 18).

The presence of organic matter distinguishes soil from other weathering products which may have accumulated above the solid rock surface. The soil profile must be distinguished from the weathering profile (Fig. 16); it is a further modification of any part of the weathering profile which occurs near, or at, the ground surface (Ruxton and Berry, 1957).

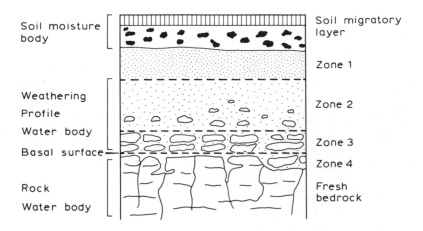

16 *Soil profile and weathering profile. Zone 1 represents the C horizon of the soil profile. Zones 2 and 3 are the weathering profile. (After Ruxton and Berry, 1957).*

The combined mantle of fragmented rock and organic material is referred to as the regolith. Variations in the thickness and other properties of the regolith are of great significance for the evolution of landforms.

In humid tropical regions, the regolith may exceed 30 m depth in places, providing a large zone of weathered material through which water may move and translocate material. Two important zones of water movement in the regolith are the soil moisture zone within the soil profile proper and the deep zone of water movement along the contact between weathered and unweathered rock (shown in Figure 16 as the basal surface of weathering). Occasionally road cuttings through deeply weathered material reveal seepage of water along the weathered rock-solid rock contact, evidence of this deep circulation, but in other exposures the seepage of moisture from a position in or at the base of the soil profile may be more prominent. Rochefort and Tricart (1959) argue that deep circulation of water along the regolith-solid rock contact or in weathered cracks and joints in the solid rock supplies most of the base flow of rivers in some seasonal intertropical climates, such as the Rio Paraiba do Sul in Brazil. Such water at the base of the regolith, out of reach of evaporative forces and plant roots, represents a moisture store.

The nature of the regolith in any one climatic situation varies as a function of lithology, slope, and position in relation to the general topography. In humid tropical Johor, West Malaysia, Swan (1970a) found that in areas of undulating relief, the variations in the texture of the upper layers of the regolith were primarily related to lithology, with position on the slope profile a secondary factor affecting the percentage of clay-silt material in the upper layers of the soil profile.

Although development of soils as a function of parent material, climate, topography, organic activity and time varies widely at the world scale, two basic patterns of soil evolution may be recognised in humid regions. Strakhov (1967) distinguishes the podsolisation of the cool temperate, taiga, forest zone from the laterisation of the tropical forest zone. The coniferous trees of the taiga do not require the bases (Ca^{++}, Na^+, K^+) and thus do not restore them to the soil surface. As a result, humic acids, derived from the decaying vegetable matter of the forest floor, leach the upper soil strongly of bases, colloids, and iron and aluminium oxides, leaving a characteristic ash-grey A_2 soil horizon composed largely of silica. Colloids, humus and iron oxides carried out of the A_2 horizon accumulate in the B horizon, which may be dark in colour, dense in structure, and in some cases hardened into a rock-like iron pan.

Laterisation occurs where weathering is intense and drainage good so that nearly all the cations and silica originally combined in the primary alumino-silicate minerals are leached out of the soil profile, leaving kaolinitic clays and residual iron or aluminium oxides. Although this process is widespread in tropical rain forest areas, there is extensive evidence that it occurs in wet, well-drained areas well outside the tropics (Stephens, 1971; Paton and Williams, 1972). Laterisation develops thick B horizons of hydrous iron or aluminium oxides with leached clays above, and only a thin layer of organic debris in the A_0 horizon. Often hard nodules of iron oxide or bauxitic material are scattered through the B horizon, sometimes forming a virtually continuous layer. Such iron accumulations tend to form in virtually all parts of the landscape, the conditions for their development being a function of parent material, drainage, rainfall and to a lesser extent temperature and time.

When the iron rich B horizon is exposed to drying conditions, it tends to harden, eventually forming an indurated layer, properly called laterite, but forming part of the wider group of hardened surficial crusts, collectively known as duricrust (Woolnough, 1927). Laterite is a specific type of the iron oxide group of duricrusts known as ferricrete. Bauxite, the product of the hardening of gibbsite-rich

B-horizons, should be regarded as aluminium-rich duricrust. Silica-rich indurated horizons are widely known as silcrete, another variety of duricrust.

The usage of such terms as laterite, ferricrete, bauxite, weathering crust and evaporation crust may be clarified by classifying duri-crusts in terms of bulk chemical composition, that is the propor-tions of SiO_2, H_2O_3 and Fe_2O_3 (Dury 1969a) (Fig. 17). The ternary diagram does not include those duricrusts dominated by calcium, gypsum and sodium chloride found in arid areas.

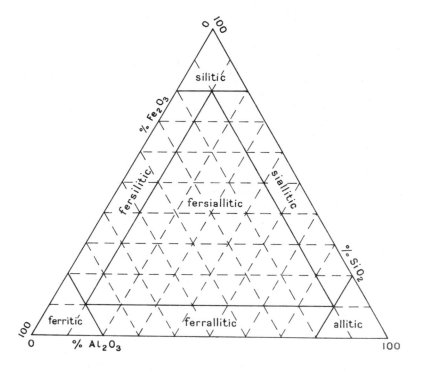

17   *Ternary diagram suggesting boundaries among the seven types of duricrust in the total fersiallitic range (after Dury, 1969a).*

Soils, duricrusts and time

Many of the soils and duricrusts found today are not developing under present-day conditions but were formed when environmental conditions were different from the present. As time proceeds, soil material becomes more weathered and soil profile horizons become more differentiated. On hillslopes near Nowra, New South Wales, three groups of soils can be found. The oldest soils are characterised by deep red and yellow podzolic soils with well differentiated profiles, while grey-brown soils without marked contrasts between horizons form a second group. Thin, poorly organised dark soils represent the youngest group, little soil evolution having occurred on the parent material. Carbon 14 dating suggests that the oldest group may be more than 20,000 years old, the middle group about 4000 and the youngest soils formed during the last few hundred years (Butler, 1967).

In addition to gross profile characteristics, the clay mineralogy and micro-morphology provide clues in the elucidation of environmental change and the duration of soil evolution. The percentage of clay in the soil and the proportions of clay minerals of different types in a soil, either at the surface, or buried by later deposition, indicate the efficiency of the weathering environment in which that soil evolved.

Although duricrusts may occupy only a small fraction of the total land surfaces in which podsolisation and laterisation prevail, they have a significant effect on landform evolution. In east-central West Malaysia, ferritic duricrusts form hard cappings to ridges on sedimentary rocks, protecting the underlying rocks against erosion. As such duricrusts appear to have evolved under environmental conditions which no longer exist, they must be regarded as relict features, representing previous stages of landform evolution. The widespread occurrence of ferritic and siliceous duricrusts in Australia, including the humid area of the New South Wales coast, has been taken to indicate a long period of warm humid conditions which favoured deep weathering and the accumulation of iron, presumably as a residual material following the loss of more soluble elements in solution. However, as the influence of lithology and relief on duricrust formation may offset that of climate, residual ferritic duricrusts of the type found in Australia are poor indicators of past temperature conditions (Paton and Williams, 1972) although their presence in dry areas may be taken as an indicator of more humid climatic conditions in the past.

With such relict features occurring in equatorial areas, their presence in most humid climato-morphogenetic environments is to be expected. As the alternations through time of geomorphic systems in areas that are presently forested become more clearly recognised, the roles of past phases of weathering and soil formation and of duricrusts in the evolution of the landforms of humid areas will emerge with greater clarity.

The study of the details of the fabric of the soil, the arrangement of grains, soil particles (peds) and the layers of clay surrounding larger grains (clay skins or cutans) reveals whether or not certain soil constituents have been concentrated or re-arranged. Micromorphological investigations also consider the chemical and physical processes and conditions under which these concentrations and re-arrangements have occurred (Brewer. 1972). Often soil fabric features which would not have formed under present conditions are recognised as evidence of climatic change (Bos, Jungerius and Wiggers, 1971). Such interpretations, although subject to varying amounts of error, can add to the recognition and relative chronology of changes in the operation of denudation systems.

In all discussions of the influence of soil formation on landform evolution, the influence of topography on pedogenesis has to be remembered. The formation of ferricrete, for example, depends, in some situations, on the lateral downslope migration of elements in solution, with an accumulation of iron in topographic lows, particularly at the base of valley-side slopes (Lamotte and Rougerie, 1962).

Weathering and soil formation thus represent the start of the mobilisation of elements which migrate as solid particles, in solution or in organic complexes.

IV

WORK OF WATER ON SLOPES

In well-vegetated terrain, plants so modify water action on slopes that their role in interception and infiltration requires further elaboration before the effects of water on soils and rock materials are discussed.

INTERCEPTION

Interception influences the work of water on slopes because it breaks the fall of raindrops and slows down the rate at which precipitation reaches the ground surface. Salts left on leaf surfaces through the evaporation of intercepted rain will be washed from the forest canopy by later rains (Stanhill, 1970), so increasing the concentration of dissolved solids in water coming into contact with soil and rock particles. Interception thus increases heterogeneity in the distribution of soil water and associated soil nutrients, so giving variations in moisture and weathering conditions over short distances.

The proportion of rainfall which is intercepted varies greatly between different forest formations and in time at any given site. Variations in the amount of throughfall over short distances in tropical forests which persist for several years arise from the spatial distribution of shoots in the canopy and are independent of the characteristics of individual storms (Hopkins, 1965). Similar effects occur in temperate forests (Ovington, 1964), German beech forests intercepting only 8 per cent of annual rainfall while German spruce forests intercept 26 per cent (Eidmann, 1962), New South Wales radiata pine plantations 18·7 per cent and New South Wales eucalypts 10·6 per cent (Smith, 1974). Seasonal leaf fall probably accounts for the low interception by beech and other deciduous trees.

Despite the wealth of research on the influence of leaf form on interception, Pereira (1973) argues that interception is a physical process governed more by the magnitude and duration of the storm than the biological character of the foliage. A comparison of rainfall interception over a range of storm sizes for various tropical and temperate forests (Fig. 18) indicates that the hydrological importance of interception by a continuous forest canopy depends directly on the pattern of rainstorms and is similar for many species of trees; however, for light falls, interception by eucalypt forest is markedly lower than for other tree species.

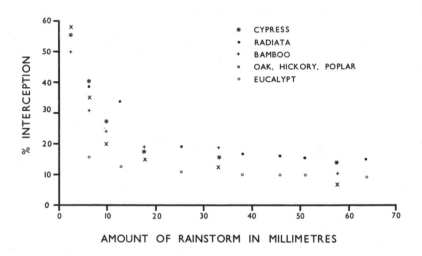

AMOUNT OF RAINSTORM IN MILLIMETRES

18 *Rainfall interception over a range of storm sizes for various tropical and temperate forest types (based on data by Pereira (1973) and Smith (1974)).*

INFILTRATION

Once water reaches the soil surface it wets the surface cover of the soil, and when this wetting is completed, extra water either penetrates the surface layers or if the surface is temporarily or locally impermeable, runs off towards the stream channel. If the surface layers of the soil are porous with minute voids for water droplet passage into the soil, the water infiltrates into the subsurface soil. Once water has infiltrated through the surface layers, it may percolate downward under the influence of gravity until it reaches the zone of saturation at the phreatic surface.

    The actual rate at which water enters the soil is termed the infiltration rate and is a function of the rate of supply of water to the soil surface. The maximum rate of water entry into a soil is termed the

TABLE 4 **Infiltration capacities of soils in humid areas**

Soil type or texture	Location	Infiltration capacity mm/hr	Source
Temperate zone soils			
Arable soils: clay loam	U.S.A.	2.5– 5.0	Morgan, 1969
silt loam	U.S.A.	7.5–15.0	,,
loam	U.S.A.	12.5–25.0	,,
loamy sand	U.S.A.	25.0–50.0	,,
Pumice soils with *Pinus radiata* forest	New Zealand	630	Selby, 1967a
Pumice soils with immature *Pinus radiata* (3–2 m high)	New Zealand	47	,,
Tropical forest soils Batu Anam soils (lateritic silty clay)	Malaysia	10.0–13.6	Eyles, 1967
Durian soils (lateritic silty clay)	Malaysia	8.6–89.0	,,
Malacca soils (massive laterite)	Malaysia	15–420	,,
Malacca soils (massive laterite)	Malaysia	107–650	Tippetts and others, 1967
Segamat soils (reddish brown latosols)	Malaysia	116–132	,,
Deeply weathered gneiss *Imperata cylindrica* cover	Ibadan, Nigeria	84–240	Thomas, 1974
Savanna soils Sandy terrace latosols	Rupununi, Guyana	233	Eden, 1964
Sheetwash latosols	Rupununi, Guyana	70	,,

infiltration capacity and is determined by soil and associated properties. The infiltration capacity is measured by artificial application of water to the soil surface at known rates. Such experiments
(Table 4) indicate that the infiltration capacities of forest soils are
sufficient to permit infiltration of all precipitation save that from
the most intense and infrequent severe storms. As many soils have a
layered structure, measurements of infiltration at the surface may

be inadequate to express the overall rate at which water may enter the soil. Understanding of the relationship between the soil type and infiltration characteristics is needed for studies of runoff and erosion (Turner, 1963).

Much more water infiltrates into mature forest soils than into bare or disturbed soils (e.g. the New Zealand pumice soils in Table 4). In Java 90 per cent of approximately 60 mm of rain which fell in 24 hours was retained by forested soils on recent volcanic rocks, but only 40 per cent of the same fall was retained by adjacent deforested soils and 65 per cent by cleared and terraced soils (Bakker and Van Wijk, 1940). Rainfall duration and intensity, as well as soil and plant characteristics, affect infiltration and moisture retention.

Much of the retained water is used by forest vegetation. Transmission of water through soil to subsurface aquifers under *Pinus radiata* forest in south-eastern South Australia is less than half that under adjacent pasture (Colville and Holmes, 1972) while streams draining the forest of the Rio Doce area of North Minas Geraes, Brazil, are waterless in the dry season, although in adjacent cleared areas they contain water throughout the year. Such reduction of runoff through use of water by trees may be less marked in more rugged country where deep weathering prevails. Here the infiltrated soil water may seep beneath the root zone, giving outflow to rivers throughout dry periods, but causing trees to cease transpiration and close their stomata.

SOIL MOISTURE AND SOIL PROPERTIES

In a uniform soil profile when there is no water arriving at the soil surface, water will be held above the level of the water table by surface tension on soil grains. Immediately above the water table water forms continuous moisture films with entrapped air pockets, films becoming less continuous further away from the water table. This zone of water tension above the water table is termed the capillary fringe. Above the capillary fringe moisture is held in discrete menisci between grains.

When infiltration begins, water passes down into the soil as a descending wetting front under the influence of gravity. Surface tension and capillary suction will oppose this descending front and thus water may take an irregular distribution within the profile, with layers of wetter soil lying between drier layers above the water table. Where the material within the soil profile is heterogeneous the water distribution may be even more complicated. On slopes, water movement is laterally downslope, as well as vertical.

Soil moisture movement affects soil formation and slope stability, both the shear stresses, the forces acting tangentially on particles, and the shear strength, the resistance to the tangential forces of soil and regolith materials being partly dependent on it. As surface materials become completely saturated, the development of positive pore pressures decreases the internal friction within the soil body and thus the stability of the slope. Many natural landslides arise from this development of positive pore pressures during heavy rain, as happened on many sections of the Darling Downs escarpment in Queensland during an exceptional storm in January 1974.

Some soil mantles never attain complete saturation and never fully dry out, thus having negative pore pressures all the time, which enables their surface slopes to stand at angles which exceed the shearing resistance. Such an angle will exceed 45° in silty material where capillary suction is persistent but will be about 35° in loose scree material where pressures in large pores are close to that of the atmosphere. Nevertheless, in humid areas, the majority of hillslopes stand at angles which are less than the angle of shearing resistance of the soil mantle, and this is probably due to the development of positive pore water pressures at times of prolonged rainstorms which give rise to perched water tables due to heterogeneity in soil profile composition.

Influence of soil profile characteristics on moisture movement and slope stability

As soil evolution produces profile differentiation, most soil profiles show varying grain size and permeability characteristics in the vertical direction. Thus the development of contrasts in pore water pressures may be expected in many soils. Other contrasts in permeability would be expected between the weathered rock material and unweathered rock. In the Exmoor and Pennine districts of Britain, the weathered mantle overlies a zone of disintegrating bedrock where joints are opened by frost action and solution (Carson and Petley, 1970). Permeability decreases at the junction of the weathered mantle with the zone of fractured bedrock and decreases further at the base of the fractured zone. This second

Humid Landforms

abrupt fall in permeability at the junction between weathered and unweathered bedrock creates a perched water table which, after prolonged heavy rain, produces a saturated soil body above the bedrock and thus gives rise to mass movement of the soil body downslope.

Permeability is thus one important soil property affecting slope stability. For a given soil, permeability may be determined experimentally in the laboratory, samples being taken for each soil horizon to reveal profile variability. Theoretical values for permeability (Table 5) show that permeability of sand is a million times greater than that of clay (Tricart, 1965a).

TABLE 5 **Specific permeability of certain soil classes and U.S.A. rock types** (after Todd, 1964 and Garrels and Mackenzie, 1971)

	Specific permeability (m darcy*)
Clean gravel	$10^6 - 10^8$
Clean sands and clean sand and gravel mixtures	$10^3 - 10^6$
Woodbine sandstone (Cretaceous)	1.5×10^3
Leduc limestone (Devonian)	0.8×10^3
Smackover limestone (Jurassic)	0.74×10^3
Find sands; silts; sand, silt and clay mixtures	$10^{-1} - 10^3$
Stevens sandstone (Miocene)	1.4×10^2
Bedford sandstone (Devonian)	0.5×10^2
Spraberry silty shale (Devonian)	5×10^{-1}
Asmari limestone (Tertiary)	less than 5×10^{-1}
Unweathered clays	$10^{-3} - 10^{-1}$
Shale (Cretaceous)	4×10^{-3}
Slate (Precambrian)	1.3×10^{-3}
Chert (Precambrian)	1.9×10^{-4}

*1 m darcy $= 10^{-3}$ darcys. A permeability of 1 darcy is possessed by a material that transmits 1 cm^3 of water in 1 sec through an area of 1 cm^2 down a length of 1 cm under a pressure gradient of 1 atm.

Permeability values assist in determining when rainfall intensity may exceed infiltration, how much lateral seepage of water may be expected in the soil and when a given soil layer may become so full of water that it may pass the limit of plasticity and then that of liquidity. The mechanical properties of a clay change as a function of its water content. When dry it behaves as a solid; as water is added, it is softened and becomes plastic. These changes in state are expressed by various measures termed Atterberg limits after the Swedish soil mechanics engineer who first used them in 1905.

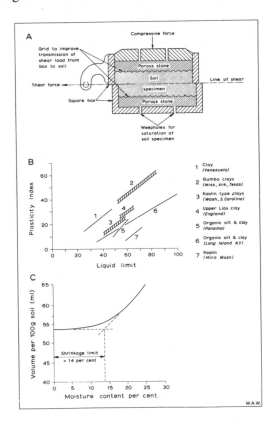

19   *Aspects of the mechanical properties of soils:*
     (a)   Diagram of a shear box
     (b)   Relation of liquid limit to plasticity index (after Kaye, 1950)
     (c)   Shrinkage limit: relationship between the volume of soil and its moisture content

Atterberg limits

The limit of liquidity is the water content at which the properties of a liquid appear in the clay. Once a soil mass has passed this limit, it will flow like a liquid without breaking into discrete units. Each particle of clay is separated from another by a film of water. Thus a clay past the liquid limit behaves as a mudflow.

The limit of plasticity involves a lower water content. The soil is mobile, but breaks apart easily. It is determined by rolling the clay into thin lengths, which break as they get thinner. The limit is passed at the water content when the clay length breaks at a diameter of 3 mm (Tricart, 1965a). The difference between the plastic and liquid limits is the plasticity index. The larger the plasticity index of a soil, the greater is its plasticity, its compressibility, and its dry strength (Fig. 19b) (Kaye, 1950).

The shrinkage limit is reached at the water content at which the clay sample has achieved its minimum volume. As the soil is dried, absorbed water evaporates and cracks appear in the clay, air then enters the soil and the volume decrease becomes appreciably less than volume of water lost (Fig. 19c).

Slipping of weathered material may occur when its water content reaches the liquid, or even the plastic limit in some circumstances. As these limits will vary with depth through the weathering profile, changes in either water table levels or the volume of infiltrating water may cause the limits to be reached at some depth below the surface and failure may occur at that point. As mass movement occurs to achieve an adjustment of conditions, materials that have moved would be expected to acquire higher plastic and liquid limits than the undisturbed regolith.

Shearing resistance

Rock and soil slopes fail when the shearing stress reaches a critical value referred to as the shear strength. Shear strength for mixed soils is generally determined by Coulomb's Law which may be expressed as:

$$S_* = C_* + \sigma \tan \phi$$

where S_* = shear strength
C_* = cohesion or cohesiveness
σ = pressure normal to the shear plane
ϕ = angle of shearing resistance or internal friction
$\tan \phi$ = coefficient of internal friction

In material consisting of discrete independent particles such as non-cemented silts, sands and larger fragments, the shear strength is controlled by the pressure normal to the shear plane and the angle of internal friction, while for cohesive materials such as clays, the shear strength is controlled by the cohesion. In mixed materials, the most usual case, all the factors are involved.

Cohesiveness, the shear resistance under no pressure, is a measure of the material's ability to adhere together. It will be negligible in non-cemented material devoid of clay minerals but will increase as the proportion of clay-sized material increases. Moisture lowers the cohesiveness of clays, and hence their shear strength, by varying amounts depending on the mineralogy of the clays involved.

In discussing the stability of slopes containing clays, the shear strength to be considered is the residual shear strength (Skempton, 1964). Most natural soils and rocks are subject to strain softening, which means that less stress is required to produce additional strain once the peak shear strength has been passed. This strain softening is not without limit. Ultimately a residual shear strength is reached which does not decrease even with large displacements (Fig. 20). Residual and peak shear strength are conveniently determined using a shear box apparatus in which the internal cohesion of a sample is measured (Fig. 19a) (Cullen and Donald, 1971).

The ultimate stable slope angle occurs when the shear strength has fallen to its lowest value, the residual shear strength. Residual strength is largely related to the mineral composition of the clay on the failure surface (Kenney, 1967) and. in experimental circumstances, more or less independent of the initial moisture content of the sample.

These interrelated mechanical properties provide a means of examining experimentally the relationship between theoretical patterns of soil movement and the actual pattern of slope development in the field. They all reflect aspects of the behaviour of water in the soil. Movement of the soil in the mass is only one aspect of the work of water on slopes; translocation of material by surface erosion and subsurface flow must also be considered.

SURFACE EROSION

Soil moisture and soil erodibility

The amount of water in a soil is one of the characteristics affecting

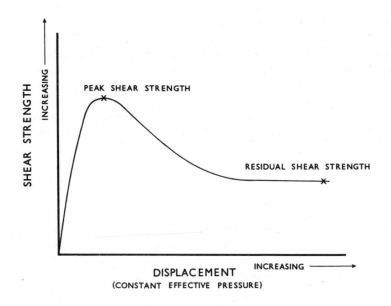

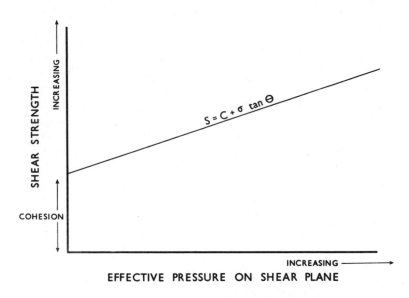

20 *Residual shear strength diagram (after Cullen and Donald, 1971)*

the erodibility of that soil. Yamamoto and Anderson (1967) use the term 'erodibility' in the restricted sense of the physical characteristics of the soil which affect its resistance to erosion. They see two additional factors, 'potential erosivity' and 'cover protectivity', affecting the actual amount of soil erosion which may occur. Potential erosivity as used by Cook (1936) includes the total quantity and rate of rainfall, the velocities of raindrops, the infiltration and storage capacities of the soil, and the steepness and length of the slope. To express the way in which a soil may respond to erosive factors, many different measures have been tried. In their Hawaii study, Yamamoto and Anderson (1967) developed indices of soil erodibility by determining in the laboratory the suspension per cent and the distribution of sizes of water-stable aggregates in each soil sampled. Suspension per cent is a measure of easily dispersed silt and clay in the soil. It is a logical measure to use, as material which is easily taken into suspension in water is easily washed away by erosive agents. The number of small water-stable aggregates in a soil reflects the manner in which a soil breaks up when wetted. Soils in which the small aggregates remain intact are more likely to suffer erosion than those in which aggregates become cohesive on wetting. The Hawaiian study using these indices showed that volcanic ash soils may be twice as erodible as soils derived from basalt or colluvium.

Erosion in tropical rain forest

Under tropical rain forest, the leaf litter is much less dense than under the seasonal rhythm of a temperate deciduous forest. In the rain forest, leaf fall is continuous, but the rate of destruction of litter is high. In the temperate deciduous forest, leaf fall occurs before the onset of a cold season when organic activity is much reduced, and thus the leaf litter persists throughout the period when the protection afforded by a growing leaf canopy is absent. This relative poverty of leaf litter under tropical rain forest means that the obstacles to water movement over the forest floor are fewer than those in temperate forest and thus, despite the great number of channels for infiltration provided by roots and animal and insect burrows, there may be more surface erosion under tropical rain forest than under temperate forests. Although rain forest margins and regrowth areas often have closed canopies, impenetrable undergrowth, and a thick layer of vegetable debris on the ground, mature primary rain forest has many canopy openings, a sparse shrub layer and a thin leaf litter (1 to 3 cm thick) (Ruxton, 1967) (Plate 1). Under such conditions, surface runoff may be greater than under temperate forest cover.

Two processes remove material from the soil surface under tropical rain forest, splash effects by raindrops and unconcentrated wash. As raindrops strike a bare soil surface they break down soil aggregates, detach soil particles and form muddy suspensions. The movement of particles and the consolidation of the soil surface by raindrop impact may cause a sealing of the soil surface and a reduction in the infiltration rate. The reduced infiltration increases the amount of water running over the surface and may lead to localised unconcentrated wash.

Raindrop splash

Studies in the eastern United States (Wischmeier and Smith, 1958) showed that an index consisting of the product of rainfall energy with the maximum 30-minute intensity of the storm is the most important measurable precipitation characteristic to explain variations in surface wash erosion of soils from experimental plots:

$$r = \frac{6.65 \ E_* I_{30}}{100}$$

where E_* is the kinetic energy of rainfall in J m^3 and I_{30} the maximum 30-minute rainfall intensity in mm hr^{-1}. The product of E_* and I_{30} is multiplied by 6·65 for each storm, to convert to the imperial units originally used, and values for all the significant storms are summed for the year. Characteristic values of the kinetic energy of rainfall for the Kelang catchment in West Malaysia calculated using a technique developed in West Africa (Charreau, 1969) are shown in Figure 21.

Constant dripping from leaves of forest trees may give rise to important slope wash effects. In the Safia-Pongani area of northern Papua, earth pillars found beneath canopy openings on bare unconsolidated soil exposed by tree fall and on slopes with little or no leaf litter indicated the role of raindrop impact (Ruxton, 1967). Pillars on bare soil patches beneath closed canopies were smaller. As raindrops from leaf drip may fall freely over 8 m or more, they are probably close to their terminal velocities when they strike the ground, thereby making an important contribution to surface soil

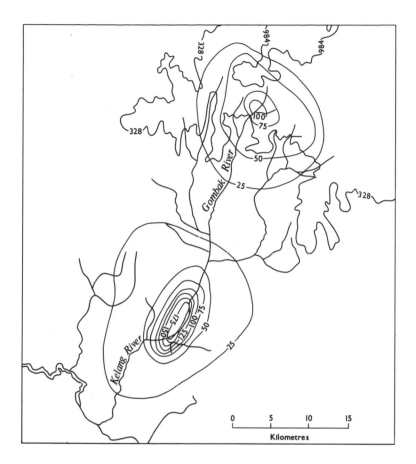

21 *Spatial variation in kinetic energy for storms over the Kelang catchment, Malaysia*

detachment in tropical rain forests. Waterdrop erosion of this kind is much more marked in tropical than temperate forests, because of the higher trees, the greater and more intense rainfall, and the less abundant leaf litter and organic debris on the forest floor.

Unconcentrated wash

Once on the forest floor, water moves downslope as un-concentrated wash, carrying with it fine particles of soil, loose leaf litter and organic debris. Root barriers and tree trunks a short distance further downslope usually trap such material (Plate 1), tree trunks in both temperate and tropical forests commonly, but by no means universally, thereby acquiring an accumulation of debris on the upslope side and a bare patch of soil on the downslope side where stemflow has washed the soil surface clear of loose debris (Fig. 9). In extreme cases unconcentrated wash may wash away the finer debris until only the coarser sand and pebbles are left to form accumulations of lag gravel such as occur locally in northern Papua (Ruxton, 1967) but not in southern Papuan forests (Ruxton, 1969). Although not always as significant in the total erosion of tropical rain forest slopes as Ruxton asserts, unconcentrated wash occurs in areas as diverse as Bougainville (Speight, 1967a) and the Ivory Coast (Rougerie, 1960). Erosion by such wash is more efficient on steeper slopes (Demangeot, 1969) and where infiltration rates are low (Speight, 1967a).

In montane tropical rain forest areas, the ground cover is denser than at lower altitudes. Above 1500 m on the granite hills of the Main Range of West Malaysia, a thick spongy layer of black peat overlies the mineral soil (Whitmore and Burnham, 1969). Although the rate of organic matter production by the montane forests which prevail at this altitude is much less than that of the lowland rain forest, the organic debris is mineralised at an even slower rate, taking several hundred years to decompose compared with a few months in the lowland forests. The contrasts in the action of water on slopes in highland and lowland forests produced by this difference in ground cover is well demonstrated by the findings of Bik (1967) in the highlands and Ruxton (1967) in the lowlands of New Guinea.

EROSION BY SUBSURFACE WATER

If soils are examined in thin section under a petrographic microscope, sand grains may be found to have been covered with layers of clay, carried down by infiltrating water. Such 'clayskins' around larger grains in soils are evidence of the translocation of fine particles by water moving through soils. This erosion and deposition of material by soil water may be an important component of denudational processes on slopes in forested terrains where most rainfall infiltrates.

Lateral mechanical eluviation

The persistence of three sizes of material, clay, sand and residual core boulders, in granite weathering mantles gives the regolith certain mechanical properties which affect its stability. Water moving through the regolith washes colloidal particles between the coarser grains by the process of lateral mechanical eluviation (Ruxton and Berry, 1961), as occurs in the headwaters of many streams draining the granites of the Malay Peninsula and offshore islands, such as the Bukit Timah catchment on Singapore (Douglas, 1967a). On Bukit Timah quartz grains are dominant in the upper layers of the regolith on the interfluves, but absent from the material immediately adjacent to first and second order stream channels. The clays adjacent to the streams appear to be almost permanently wetted, and to have accumulated by the subsurface flow of water and clay particles through the regolith beneath the forest covered slope. Similar processes occur on the slopes of both the Bartle Frere-Bellenden Ker range in north-east Queensland and the forested upper Dekale River basin on the granite Loma Mountains in Sierra Leone (Daveau, 1965). In extreme cases subsurface water removes sand and even gravel from between residual core boulders, Labertouche Cave, Neerim South in Victoria having been formed in this way.

In humid subtropical conditions in the North Island of New Zealand (Selby, 1968) and in south-east Queensland, on greenstones and phyllites in the headwaters of the Caboolture River (Arnett, 1971), water flows through surface horizons of the soil, occasionally forming small tunnels or pipes, proving that material is being removed by lateral mechanical eluviation or solution (Plate 4). Such washing of clay particles downslope by throughflow, with development in exceptional cases of piping and percolines of greater subsurface water movement than in adjacent parts of the slope, is probably the normal case in humid areas (Kirkby, 1969). Percolines are lines of diffuse subsurface water movement above the heads of permanent channels. In many cases, piping occurs in parts of the percoline network. Enlargement of pipes may lead to subsidence, producing a series of hollows which eventually develop into a gully, as happens frequently in the deeply weathered granite country of the Northern Tablelands of New South Wales.

WATER AND SLOPE INSTABILITY

Under extreme conditions, lateral mechanical eluviation can lead to the development of subsidence features. The summit areas of the

4 *Piping, or tunnelling, in the soils of a volcanic-alluvial fan on Mt Duau, Papua New Guinea (CSIRO photo)*

northern Sula Mountains in Sierra Leone are capped by a highly permeable duricrust up to 12 m thick, beneath which is a stiff impermeable clay derived from the weathering of amphibolite schists. The rainfall of 3000 mm per year is concentrated into a wet season from May to November when heavy, intense falls are expected. The infiltration of this water through the duricrust leads to the development of underground channels and subsurface removal of the clay at the upper surface of the impermeable clay layer (Thomas, 1969). An exceptional case of long caves beneath a lateritic crust has been reported from New Caledonia. This working out of the clay eventually leads to the collapse of the overlying duricrust, and to the instability of the surface layer generally, producing slumping and

TABLE 6 **Forms of mass wasting**

Type of movement	Rate of movement	Characteristics of deposits and associated landforms
Water assisted flows		
Solifluction	10 m/hr	Slow downslope of movement of debris saturated with water. Heterogeneous deposits in lobes over pre-existing soils.
Earthflow { discrete extensive }	10–100 m/hr	Discrete flows within bounding scars, extensive flows involving large sections of hillslopes. Predominantly regolith material moving over slopes forming large lobes.
Mudflow { discrete extensive }	100–1000 m/hr	Large quantities of mud, earth and rock debris flowing downslope and spreading large lobe out at base of slope usually along a defined channel
Torrential mud cascades	1 km/hr	Surge of soil and rock debris in a water stream down narrow track producing large lobe of debris at end of flow. Large blocks carried among ill sorted material.
PREDOMINANTLY GRAVITATIONAL FLOWS AND SLIDES		
Creep	1 m/yr	Almost imperceptible downslope movement of soil and rock debris. Gradual tilting of trees, etc.
Slump	Variable — often initiated abruptly with slow movement thereafter	Rotational slipping of regolith or rock debris, producing slump scars, hummocky terrain and slump toe deposits over pre-existing soil profiles
Debris slide Debris fall Rock slide Rock fall	Rapid but involving differing quantities of material from single rock particles to whole mountain sides	Rolling or sliding of material parallel to slope / Fall of debris from vertical or overhanging slope / Sliding or falling of individual rock masses down bedding, joint or fault surfaces / Free falling of rock blocks over any steep slope
Subsidence	Rapid	Downward displacement of material into subsurface voids with no lateral movement
Debris avalanche	Extremely rapid	Surge of soil and rock debris as a dry flow usually from an earlier slip scar

earthflows involving the laterite rubble and underlying clay on the steeper slopes. Similar landslip features have been produced by subsurface removal of material in the Brive district of Limousin in western France (Meynier, 1961).

The slumping of material is one form of mass wasting, the bulk transfer of masses of debris downslope under the influence of gravity (Table 6). Water is involved in nearly all mass wasting processes, either in the breakdown of rock material and consequent development of instability, or in assisting the downslope movement of material. The water-assisted movements of material range from the slow movement of saturated soil, termed solifluction, to rapid, locally catastrophic debris avalanches.

Solifluction

The slow flowing from higher to lower ground of masses of soil or earth saturated with water occurs in all humid climates, but its significance under forest cover is not fully understood. On cultivated land in Poland on the Lodz Plateau, solifluction, although limited to small areas on concave slopes, is one of the most effective erosion processes, producing characteristic loops of soil flow material on bare ploughed fields (Koziejowa, 1963). This solifluction is a seasonal process associated with the snow melt period in spring. As such it may perhaps be considered as part of the periglacial processes of mass wasting (Davies, 1969), but it occurs well outside areas considered to have a periglacial climate. In mountainous areas of the humid tropics, solifluction is common, with disturbed layers of regolith material containing weathered rock fragments flowing over grooved and striated weathered sedentary bedrock surfaces (Ruxton, 1969). In many such areas the movement of regolith material by mass wasting is so frequent that strongly weathered soils are uncommon and most slopes have colluvial soils of solifluction and slump derived regolith (Haantjens, 1970).

Earthflow and mudflow

Earthflows, which move at rates of several centimetres or metres per day according to water availability, are common on steep slopes in such humid tropical areas as Hawaii (White, 1949), New Guinea

5 *Mudflow in lower montane rain forest near Tumundan, Western Highlands
District, Papua New Guinea. The flow is a jumble of timber, soil, mud
and weathered rock. (Photo by J. N. Jennings).*

(Ruxton, 1967, 1969), and Ceylon (Cooray, 1967), where earthflows often develop from a soil slip which becomes saturated, thus causing a mixture of water, clay and sand to move downslope as a slushy mass. Earthflows associated with slumps in the South Island of New Zealand are often discrete earthflows, bounded by distinct shear lines, with a concave upper part and a convex lobed toe (Crozier, 1968). The second major form of earthflow, the extensive earthflow, may be produced by a coalescing of discrete earthflows, but in most cases it involves the mobilisation of large subsurface areas of a slope where the density of percolines is high. The movement of earthflows is closely related to climatic conditions, being seasonal, greatest in spring in Otago (Crozier, 1968), in winter in Northern Ireland (Prior *et al.*, 1971) and in all cases closely related to periods of maximum water content in the regolith. Mudflows are faster forms of earthflow, which may move at speeds of several hundred metres per hour. Mudflows contain a variety of ill-sorted, unworn fragments. They are characteristic of steep slopes, often associated with small streams whose valleys may suddenly be overloaded with debris which flows down into the major valley in the form of a large lobe of mud, rock fragments and boulders (Plate 5).

Torrential mud cascades

Torrential mud cascades are rapid mudflows which move like lava flows at several kilometres per hour (Tricart, 1957) and which occur in mountainous terrain where weather conditions cause large amounts of relatively uncohesive material to be caught in a water stream and then flow rapidly downslope. In August 1951 severely gullied gypsiferous mudstones in the Austrian Voralberg were so rapidly eroded by intense rain that whole ridges between gullies slumped into a mountain torrent, transforming the little stream into a surging mass of mud carrying boulders up to 4 m long which spread a large tongue out into the main valley (Tricart, 1957). Mudflows and debris avalanches are particularly likely to occur in unconsolidated deposits left by glacial activity or recent volcanic activity and in weak but impermeable strata overlain by massive permeable beds. Many peat bogs tend to slide down slope, Muckle Moss in northern England moving about 5 cm yr^{-1} (Moore and Bellamy, 1974). Occasionally catastrophic peat slides, or bog bursts, develop, creating another form of mudflow.

MAINLY GRAVITATIONAL MASS MOVEMENTS

Two main types of downslope movement, creep and slipping, occur under gravitational stress without a dominant role being played by water. Creep is the slow downslope movement of superficial soil or rock debris, usually imperceptible except to observations over long periods. It appears to be the result of the displacement of soil particles through the persistent effect of intergranular forces which move soil particles in relation to other soil particles, so changing the strength and stability of the soil or regolith (Culling, 1963; Trask, 1950). Such movements may be random, but are affected by gravity and seepage water and by changes in pore space with depth. These factors lead to a net downslope migration of particles. Such a process is considered characteristic of humid climates where weathering is supplying additional material to the regolith (Culling, 1963).

In humid areas two types of creep occur, seasonal and continuous. Seasonal creep, the product of seasonal fluctuations of moisture and temperature, results in the downslope movement of a sheet of soil and regolith material with a depth equal to or less than that at which seasonal fluctuations are felt. Kirkby (1967) shows that these fluctuations are a series of cyclic movements which set up forces felt over an entire hillside without any likelihood of causing more marked disturbances of a landslide type (Fig. 22b).

Most creep of significance for slope evolution is of the seasonal type, although rock creep, either in the form of cambering for nearly horizontal strata (Fig. 22a) or of downslope rock creep for almost vertically aligned sediments (Hills, 1940) may be a consequence of superficial movement which has effects on deeper material. When beds are tilted towards the vertical, rock creep of the exposed edges of the strata may produce a curve of the rock in a downslope direction known as terminal curvature. In gently inclined strata, such as the Jurassic and Liassic limestone, sandstone, siltstone and clay sequences of western Europe, cambering is widespread. Cambering (Fig. 22a) results from a lowering of the outcrops of hard rocks capping hills and ridges; the hard beds develop curved bases and are inclined to varying amounts down the valley sides. In central England, the valley-ward creep of the Marlstone Rock bed has been produced by both outward squeezing of the underlying silty clays and by actual downhill slipping of the bed itself (Edmonds, Poole and Wilson, 1965). In deeply weathered tropical soil mantles, the

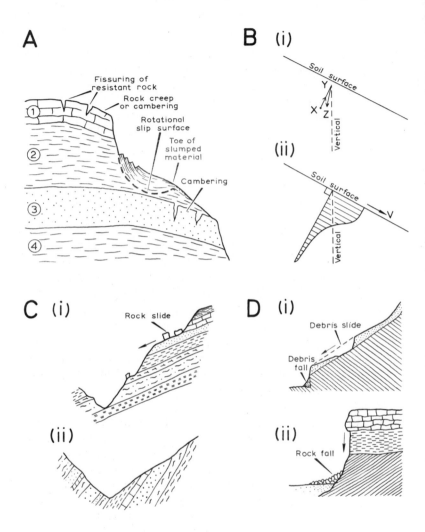

22   *Aspects of mass movement:*

    A   *Cambering in resistant strata and slumping of weaker strata typical of Mesozoic terrains of western Europe:*
        1)   *Sandy limestone caprock cambered towards valley,*
        2)   *Clay horizon affected by slumping,*
        3)   *Calcareous sandstone cambered towards valley,*
        4)   *Clay horizon*

    B   *Theoretical aspects of soil creep:*
        (i)   *Movement of an individual particle from X to Y then Z,*
        (ii) *Velocity profile for soil creep; velocity increases in the direction V,*

C   Sketches showing the stability of slopes:
    (i) Unstable, ground slope greater than or equal to dip of rocks,
         rock sliding in progress,
    (ii) Stable, ground slope less than dip of rocks or opposite it
D   Diagrams illustrating gravitational mass movement:
    (i) Debris slide of regolith material sliding over unweathered
         bed rock,
    (ii) Rock fall from overhanging free rock face.

presence of a 'stone-line' of quartz fragments derived from a resistant quartz vein upslope has long been considered evidence of creep under rain forest (Birot, 1966).

Observations of creep, made by studying the deformation of buried plastic tubes, wires, or rows of posts, show that in Britain seasonal soil creep carries from $0 \cdot 5$ to over $7 \cdot 5$ cm^3 of soil past a line 1 cm wide across the slope in a year (Young, 1960, 1963; Kirkby, 1967). Creep in these formerly glaciated upland sites is the dominant form of movement of material on slopes, carrying twenty times that removed by unconcentrated wash. Under forest in the Vosges, France, it is also probably the dominant form of erosion (Rougerie, 1965), while the existence of a migratory layer of rock waste on sedentary weathered rock and the tilting of tree trunks or fence posts away from the vertical have been taken as evidence of the importance of creep on tropical rain forest covered slopes (Berry and Ruxton, 1961; Swan, 1970a). Not all leaning trees must be considered evidence of soil creep (Phipps, 1974); many other factors can affect plant growth. Some of these features may be produced by solifluction following heavy rain rather than by soil creep in the strict sense, it being impossible to separate the two types of movement in such wet environments.

Slipping in humid environments

Slipping of debris involves a wide range of rapid slope movements ranging in size from small slumps to massive landslides, either of debris or of debris plus rock. Slumping is usually in the form of rotational slipping, in which movement, or slope failure, is preceded by tension cracks near the top of the slope, and then, later, sliding occurs along a concave surface (Cleaves, 1950). Such slipping may be regarded as the product of some addition to the weight of soil liable to slip overcoming the shearing resistance of the soil.

Water entering cracks or pore spaces in cohesive material can trigger off slumps, water entry being favoured by the development of deep desiccation cracks during dry periods in some seasonally wet environments. The occurrence of slips in the Wollongong area of New South Wales is closely related to periods of heavy rain (Bowman, 1972).

Deeply weathered material, swelling clays, or variable and easily broken sedimentary material such as greywacke, favour the development of slump features. Such slumps are widespread on weathered greywacke in South Auckland, New Zealand (Selby, 1966; Pain, 1969), where shear planes occur in the lower parts of the soil profile, either in the C horizon or at the junction of the B and C horizons. In the humid tropics, slumping is common in areas of moderate relief where regolith depth is sufficient for rotational movement to occur entirely within the weathered mantle. Thus often sharp slump headwalls, 5 to 20 m high with an angle of 70 to 80° are found above a mass of irregularly sorted slipped material (Tricart, 1965c). Such slumps often bury soils further down slope thus producing layers of soils and slump debris illustrating alternating stable and unstable phases of slope evolution (Bik, 1967).

Debris and rock slides

The sliding of material along a contact zone, for example between weathered mantle and parent rock, is common in many areas where slumping occurs. In the greywacke areas of the North Island of New Zealand, regolith material slides along the contact with the weathered greywacke (Selby, 1966, 1967b, 1967c; Pain, 1969) (Fig. 22d). Debris slides of this type are particularly likely to occur where bedding planes lie parallel to the slope. Slides (Fig. 22c) occur when underlying strata become so lubricated with water that they can no longer support the material above. Ofomata (1966) has described a debris and rock slide at Awgu in Nigeria where a road cutting had exposed a shelly limestone and clay band beneath sandstones. After heavy rain, the upper weathered debris and sandstone slides over the underlying shales. In the humid tropics, many such debris and rock slides occur as the result of human activity, being particularly frequent in areas of urban development (de Meis and da Silva, 1968). Rock slides occurred in January 1966 in a locality near Rio de Janeiro where joints were inclined parallel to the slope, while debris slides occurred where lines of subsurface water movement converged to lubricate the contact between regolith and rock escarpment. Whether such slides occur as frequently as slumping under natural vegetation is uncertain. In New Zealand, Pain (1969) found that flowage forms of mass wasting were usual under forest, while sliding occurred on cleared land, where mass movement was more frequent altogether.

Tectonics and landslides

While high intensity rainfalls trigger off many of the rapid mass movements, earthquakes are a potent factor in slope instability. In tectonically unstable areas like New Guinea, earthquakes result in massive rock and debris movements (Plate 6). The Torricelli Mountains were affected by force 7 and 7·9 (Richter scale) earthquakes in 1935, after which such large quantities of material were transported by landslides that the depth lost over the affected area was between 0·1 and 0·2 m. The effects of the earthquake on landslides varied as a function of the log of the distance from the assumed earthquake epicentre (Simonett, 1967). The quantity of material removed from slopes by such earthquake triggered landslides far exceeds that which would be removed by the work of water alone on the slopes. Simonett (1967) estimated that denudation in the Bewani Mountains of New Guinea, not affected by the 1935 earthquake, is about 22 cm/1000 years, while in the earthquake affected area of the Torricelli Mountains, the rate is probably closer to 100 cm/1000 years. The Madang earthquake of 1970 caused the removal of 27·6 × 10^6 m^3 of material from the slopes of the Adelbert Range of New Guinea, with a long term denudation rate of the area of the order of 100 cm/1000 years, 60 to 70 per cent of which is caused by earthquake initiated mass movements (Pain and Bowler, 1973).

Earthquakes in Assam in 1897 and 1950 caused large scale landslipping throughout the state, the relatively small interval between these large shocks resulting in most of the slope topography being the result of landslipping slightly modified by water erosion. With such violent tectonic activity and the highest rainfalls of the world, slopes in the Shillong district of Assam are among the most rapidly changing of any humid region.

Rockfalls

In a rockfall the mass in motion travels mostly through air with little interaction between the particles involved. In the Hawkesbury

6   *Extensive mass movements in rain forest near Wau, Papua New Guinea
(CSIRO photo)*

Sandstone of the Illawarra escarpment, New South Wales, joints
and cracks have opened up to a maximum of 0·6 m parallel to the
scarp and 16 m behind it. Some of these joints are partly covered by
wedges of soil, but water enters the joints, both accelerating the
decomposition of the sandstone matrix and increasing the hydrostatic pressure in the joints. Such actions loosen joints until lumps
of rock fall on to the cliff-foot apron below. At the Dombarton
Siding slide on the Illawarra escarpment, the rock debris consists
of unsorted broken blocks from 5 m in diameter down in a sandy
clayey matrix (Bowman, 1972). While the bare rock has a slope of
40°, the debris has an inclination of 35°, at the upper end of the
range of the repose for joint-block debris.

Slumping and subsidence

Although the term slumping is often used loosely for a variety
of types of mass movement, it is more strictly defined as the bulk
transfer of masses of debris down slope under the influence

of gravity, such as the collapse of lateritic layers described earlier. Subsidence is a special case of slumping where the fall of debris is vertical and is usually the result of removal of material from beneath a resistant layer and the eventual collapse of that layer. In addition to the effects of lateral mechanical eluviation beneath duricrusts, subsidence can occur wherever large subsurface voids are developed.

Debris avalanche

Many of the landslips triggered by earthquakes involve a combination of slumps, slides and rockfalls in which a large mass of material falls down the valley side extremely rapidly under the influence of gravity. Such slips are debris avalanches and usually occur in tectonically unstable mountainous regions such as the Andes, Himalayas and New Guinea (Plate 7). Debris avalanches may set up minor earth tremors which would be recorded as minor seismic shocks.

THE SLOPE EROSION SYSTEM

Any slope may undergo some components of the various forms of erosive activity mentioned above, from raindrop splash to landslipping. Although biological activity and creep may be virtually continuous, much erosive activity is episodic in occurrence, from the rare earthquake event of great magnitude, to the impact of raindrops from irregularly spaced rainstorms of contrasting intensity. At different times, water will be distributed over and into the surface layers of the slope in various ways, thus provoking changes in the relative contribution of surface erosion, subsurface erosion, and mass wasting to the total denudation of the slope. In order to accommodate these changing conditions of slope erosion in a general denudation system, a subsection of the system may be suggested for slope erosion. In the slope erosion system, the processes that change the slope may be considered to be the interaction between a set of external variables relating to the climate and associated biological activity and a set of internal variables of slope geology, particularly lithology and structure, but including tectonic activity. These two sets of variables operate between two end points, the upper end of the slope, which may be in a hard caprock, plateau surface or ridge crest, and the base of the slope, which may be a

7   *Debris avalanche in Gembola Valley, Upper Chimbu, Papua New Guinea*
    (CSIRO photo)

zone of unlimited waste removal, or a zone of no waste removal, or,
more normally, somewhere between these extremes. Slope forms
will alter as a result of the operation of this slope erosion system,
the rate and nature of such alterations varying as the influence of
particular components of the system changes through time.

V

HILLSLOPE FORMS

While in surveying or roadmaking slope may mean gradient, or the angle to the horizontal at which a part of the earth's surface is inclined, in landform studies slope has a wider meaning, implying any inclined, geometric element of the earth's surface. The ground surface is composed either of sloping or flat elements, several of which may combine in a single valley side slope. Slope may be considered the cardinal relief feature (Jahn, 1956) and the study of the form, significance and development of slopes has long held a nodal position in geomorphology (Chorley, 1964). As slopes are the product of the interacting factors of an ever-changing denudation system, they may also be regarded as the expression of the evolution of the denudation system up to the present, containing features derived from past erosional and depositional environments as well as features being created by the present morphoclimatic conditions.

As slopes are the expression of the way in which past and present processes have altered, removed and accumulated rock materials at the lithosphere-atmosphere interface, both the nature of the rock materials and the morphoclimatic environment would be expected to influence slope forms. As processes vary in significance with landform size, slope studies focus attention on the problem of scale. Such studies, therefore, ask both whether slope forms in humid regions differ considerably from those of other regions and whether, within humid regions, differences due to lithology and structure obscure any contrasts which may exist between slopes of different major morphoclimatic regions.

Possibly through association with gradient, slope form has usually been considered as the slope profile, the cross-section of a hillslope drawn up by measuring slope angles and distances along a line of traverse following the direction of steepest slope from ridge to valley floor, orthogonal to rectilinear contours (Pitty, 1969). However, much of the movement of water towards streams, and thus much of landform evolution, is affected by concavities and convexities along contours, at right angles to slope profile lines. The control of the convergence and divergence of debris and runoff by slope plan is a

81

critical factor in slope development (Wilson, 1968). Creep and runoff will be more diffuse downwards on a slope that is convex in plan, where the slope profile lines would diverge. Where a slope is concave in plan, the slope profile lines will converge at valley-heads and lines of water and debris movement will concentrate in the place of greatest concavity.

In the densely forested central Appalachians small first order tributaries occupy valleys with amphitheatre-like heads, steep side slopes with slightly convex-upward profiles and narrow channel-ways with little or no bottomland along the valley axes. Within such valleys, slopes were divided into five categories with which vegetation types were roughly coincident, and which also reflected the behaviour of runoff or seepage water as it moves downslope (Hack and Goodlett, 1960) (Table 7). Slope plan, as expressed by contour curvature, and slope profile often combine to create convergence towards stream head hollows and convex slopes with divergent flows at the lower ends of valley sides (Fig. 23). As com-binations of plan and profile views of slopes are difficult to express quantitatively, comparisons between slopes on different lithologies and in different climates rely on the section of the slope plan which most simply reflects the way in which denudational forces and cohesive and resisting lithological factors are adjusted to provide a steady supply of debris to the streams. The steepest part of a

TABLE 7 **Slope categories in first-order valleys, central Appalachians**
(after Hack and Goodlett, 1960)

Nose	Contours convex outward (away from mountain). Runoff proportional to a function of the radius of curvature of the contours.
Side slope	Contours straight. Runoff proportional to a linear function of slope length.
Hollow	Contours concave outward. Runoff proportional to a power function of slope length.
Channelway	Contours sharply concave outward. Runoff proportional to a power function of channel length.
Footslope	Transitional area between side slope and channelway (not present in all valleys).

valley side slope, the maximum angle which can be maintained, is a convenient indicator of the relative effectiveness of the opposing forces of the slope erosion system.

Slope profiles have been used as the basis of comparison between erosional landforms since Gilbert's explorations of the Henry Mountains of the United States and his observations on the steepness of slopes in relation to lithology, distance from divides and from stream channel.

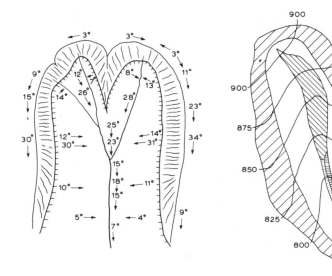

Twin dells near Szár, Hungary

Valley on west side, Crawford Mountain, U.S.A.

23 *Contour curvature and valley form. The Hungarian example uses slope angles to indicate variations in slope at the head of a small valley, while the American example divides the valley into morphological units:*

1. *Nose*   3. *Hollow*   5. *Footslope*
2. *Side Slope*   4. *Channelway*

These terms are defined in Table 7.

Scale of variation in slope topography

Slopes are full of irregularities. Every climber and fell or bush walker knows that a slope which appears smooth from a distance may be rough and boulder strewn in reality. Is this detailed roughness of greater significance than the apparent overall smoothness? If the slope is being compared with slopes in contrasted morphogenetic regions, the overall characteristics may be more important, but if it is being compared with other slopes in the same valley or same general area, then the detailed characteristics will be significant. Savigear (1967) has suggested that this problem of scale in slope studies may be tackled by recognising that a hillslope from divide to stream channel consists of a number of *components*, which are themselves composed of a number of smaller forms called *units*. The unit itself is made up of smaller *micro-units*, which in turn are subdivided into *textural units*, which range in size down to individual clay particles. Savigear's hierarchy, from largest to smallest units, is thus: Slope — component consociations — components — units — micro-units — textural units.

Erosional and depositional zones of slope components

As water removes material and carries it down a slope, it tends to erode the upper section of the slope but, if the removal of material at the base of the slope is not equal to the rate of supply of material from above, it may deposit material at the foot of the slope, thus creating an upper erosional zone and lower depositional zone on any given slope. While screes and talus zones are common in formerly glaciated areas or areas of periglacial activity, extensive slope foot debris accumulations are not common in humid areas unaffected by climatic changes. Nevertheless, the fundamental distinction between erosional and constructional slope forms is essential for understanding the nature of slope components. Fringing many hillslopes in the humid tropics are zones of colluvium, material either washed downslope by water or carried by mass movement. These colluvial deposits are particularly important in tectonically active areas like New Guinea. However, extensive colluvial deposits along the great escarpments of the humid fringe of eastern Australia were probably produced in drier periods during the Quaternary when the rate of removal of material from the slopes exceeded the rate of weathering. In the Hunter Valley area such colluvial deposits are no longer accumulating and carry mature soils (Galloway, 1963).

Even on a uniform lithology, the contrasted effects of erosion on the upper part of a slope and accumulation on the lower part would be expected to produce differences in form. The depositional and

erosional parts of slopes may be regarded as two separate components. However, the erosional part of the slope is seldom
dominated by a single process of erosion and thus other slope components may result from differences in the relative dominance of
particular processes. Many writers have commented that erosional
slopes contain an upper convexity where soil creep is dominant and
a lower concavity principally formed by concentrated flow, which
becomes more and more efficient towards the base of the slope
(Baulig, 1940). Such a convexo-concave slope is represented by the
lowest of the four schematic slope profiles in Figure 24. This slope
has a break between the convex and concave components, but as
the South Auckland ash-mantled slope shown on profile 2 in Figure
25 suggests, the convex and concave portions of such erosional
slopes may be linked by a third, straight component which possibly
lies at an angle of equilibrium between the external and internal
factors affecting slope evolution.

Simple convexo-concave slopes are but one of a variety of slope
forms to be found in humid regions (Fig. 25, 26). Lithological
variations along a slope greatly affect slope form, slopes eroded

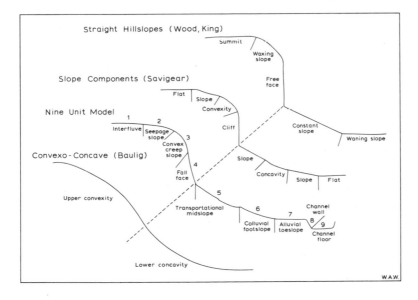

24 *Theoretical combinations of slope convexity and concavity*

SOUTH AUCKLAND   GREYWACKE LANDSCAPE

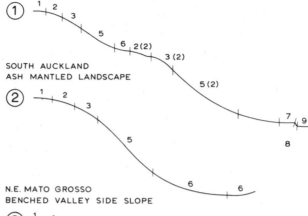

SOUTH AUCKLAND
ASH MANTLED LANDSCAPE

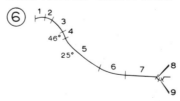

N.E. MATO GROSSO
BENCHED VALLEY SIDE SLOPE

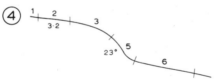

N.E. MATO GROSSO
LATERITE SCARP & RIDGE LANDSCAPE

INSELBERG, KATUNTUKWA, BUNKEYA, KATANGA

JOHOR MALAYSIA GRANITE AREAS

25 *Application of the 9-unit land surface model to slope profiles from different environments*

in weak rocks generally having gentler angles than those cut in hard resistant rocks. Many hillsides cut in the clay formations of southern England and the Paris Basin slope at angles of 10° to 15°, while slopes developed on limestones and sandstones of the same areas range from 30° to 40° (Dury, 1959). Vegetation also influences slope; loose sand and gravel debris in Connecticut was found to have an average angle of repose of 33°, while natural forested slopes on this material had a mean angle of 41°, the trees stabilising the material at a significantly steeper angle (Rahn, 1969). These lithological and biotic factors introduce complications into the simple convexo-concave type of slope. Resistant rock bands sometimes outcrop as a bare rock face, free of debris, with perhaps beneath it a straight slope, inclined at an angle of equilibrium. To accommodate this variety of slopes, Dalrymple, Blong and Conacher (1968) have developed a hypothetical 9-unit land surface model (Fig. 24). On the interfluve (unit 1) pedological processes involving the vertical movement of subsurface water are dominant. Unit 2 is a seepage zone where throughflow and thus lateral mechanical eluviation are important. The convex unit 3 is dominated by soil creep, while unit 4 is the free face of rock outcrop affected by rockfalls. Unit 5, the transport mid-slope, is similar, but not equivalent to, the straight, or constant slope mentioned previously. The dominant process is the transport of material across the unit. The deposition of colluvial material from above dominates the straight or concave unit 6, while unit 7, the equivalent of a river floodplain, is dominated by the deposition of alluvial material brought down the valley by the river. The erosive and transport actions of the river dominate the channel wall (unit 8) and the channel bed (unit 9).

This division of the land surface into 9 units has some similarity to the scheme of plane and curved units devised by Savigear (1965, 1967); however, the 9-unit model is not as concerned about the curvature or straightness of units. The 9-unit model also has much in common with Wood's slope component model (1942) and King's straight hillslope model (1953) (Fig. 24). The series of hillslopes on Figures 25 and 26 illustrate the flexibility of the 9-unit model, which ought to be applicable to any natural hillslope, even though, as profiles, 1, 11 and 12 indicate, many hillslopes are composite features in which units recur downslope.

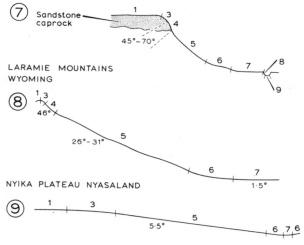

JOHOR MALAYSIA
GRANITE AREAS WITH SANDSTONE CAPROCK

LARAMIE MOUNTAINS
WYOMING

NYIKA PLATEAU NYASALAND

NEGERI SEMBILAN MALAYSIA
GRANITE LOWLAND

SAMBRE BASIN BELGIUM
SANDSTONE LANDSCAPE

SAMBRE BASIN BELGIUM
SCHIST

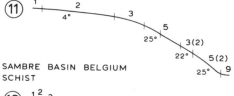
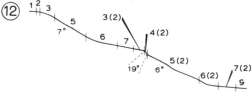

26   *Further examples of the application of the 9-unit land surface model*

CONCEPTS OF CONSTANT OR THRESHOLD SLOPES

All materials have a threshold angle of repose above which they are unstable and liable to be affected by rapid mass movements. The threshold angle of fresh unweathered debris may be quite different from the angle of a slope developed on vegetated well-weathered debris of similar lithology. Carson and Petley (1970) found that the angle of limiting stability of a waste mantle or regolith, even on a single lithological type, depended on the stage of weathering. A mantle of loose fragments may stand at 35°, which is the angle of repose of such material. A subsequent reduction of the mantle into a mixture of soil and rock fragments will increase the peak angle of shearing resistance, but also increase the pore pressures developed at times of prolonged rainstorms, so that the maximum stable slope is reduced in gradient to 25° to 26°. The production of a true soil may further decrease the angle of the stable slope to 19° to 20°. These conclusions suggest that any section of the landscape may contain straight slopes at any angle between wide limits, for example between 19° and 35° in the Exmoor area of Britain (Carson and Petley, 1970).

Although slopes may exist at characteristic angles that are less than the theoretical limiting angles (Young 1961), the characteristic angles found on a particular rock type in a given climatic region are probably related to the changes in limiting angle or threshold slope as the mechanical properties of the slope materials are changed by weathering through time. If there is such a variety of stable slope angles possible in a single environment, any notion that threshold slope angles vary with morphoclimatic conditions needs to be examined most carefully. Usually these threshold slope angle studies apply to unit 5, the transport mid-slope, of the 9-unit model. The angles of unit 5 straight slopes in the Laramie Mountains, Wyoming, range from 18° to 33° with a pronounced modal group at 25° to 28° (Carson, 1971). The modal group angles seem to be linked to the stability of the mantle in an intermediate stage of weathering. The various straight slopes indicated in Figure 27 all show a considerable range of angles. However, scree slopes tend to be grouped around a mean angle, suggesting that the scree slopes tend towards a certain equilibrium angle. But a scree slope would contain very little weathered material, and so this angle would be the threshold angle for unweathered debris. Where greater degree of weathering

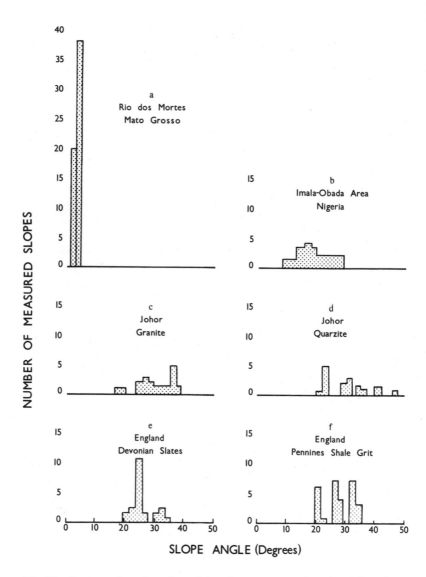

27 *Distribution of angles of straight slope segments based on data from
a) Young, 1970; b) Jéjé, 1972; c) and d) Swan, 1970b; e) and f)
Carson and Petley, 1970*

exists, the spread of the angles is greater, and the slope angle frequency distribution is far removed from the normal distribution characteristic scree slopes. Thus Strahler's view (1950) that 'within a given small region where conditions of lithology, climate, soil, vegetation and relief are uniform, slopes tend to approach a certain equilibrium angle, appropriate to those controlling factors' may require modification to include a time factor, to account for the effects of Quaternary climatic change on the weathering history which would influence the mechanical properties of the weathered mantle and therefore the equilibrium slope angle.

The discussion of threshold angles so far has been limited to the straight section of hillside slopes, unit 5, the transport mid-slope, which corresponds to the constant slope of the earlier slope models of Wood (1942) and King (1950), but Dalrymple, Blong and Conacher (1968) say that the wide range of angles occurring in this unit precludes that name. Nevertheless, the evidence on threshold angles discussed above would suggest that a wide range of slope angles may be due to differences in mechanical properties affecting the threshold angles of various materials weathered to differing extents.

As the straight unit 5 is recognised as the zone of downslope transport, with the transport across the unit being more marked than the erosion of the unit, conditions on the threshold slope may be constantly changing. Thus it is possible that a thin waste mantle may give rise to a single straight slope unit, say at 26° to 36°, while a thick waste mantle will produce two segments, at 25° to 27° for a lower unit and 32° to 34° for the upper unit for the Exmoor and Pennine areas discussed by Carson (1969). The lower segment may extend at the expense of the upper segment and gradually a gentler slope will evolve.

Whether this type of relationship between angle and process may be extended for complete hillslope profiles from divide to valley floor is not yet proven. Strahler (1950) and Pitty (1970) have used moment measures, such as skewness and kurtosis, to characterise percentage frequency distributions of slope angles over profiles. Such a technique would offer a more complete means of comparing hillslope forms, but as yet few such examinations have been made. One of the few is Melton's study of the arid environment of southern Arizona (1960), which showed that mean slope angles varied with the local erosional environment, rather than regional climate or lithology, thus implying a control of overall hillslope form by locally significant processes. In dry conditions, low channel gradients favour the development of valley asymmetry with poleward facing slopes becoming steeper as the channels and streams are moved against their toes by filling of debris from the equatorward facing slopes which have more bare soil. Such contrasts are less marked in humid regions, and should be negligible in tropical regions, but possibly significant in humid temperate regions where contrasts in insolation produce differences in frost action and snow retention.

There seems to be no reason why it should not be possible to characterise slope profiles related to specific patterns of landform evolution, but in view of the great surge of questioning of technique of slope profile measurement (Young, 1971a, b; Pitty, 1969, 1970) and of the development of process-response models (Carson, 1969; Thornes, 1971), it would not be advantageous to set out a list of characteristic slope angles at present. While slopes are of crucial significance in all landform studies, they should not be studied in isolation, but should be related to the whole morphological environment, which in humid landform areas means that slope processes and forms must be related to river processes and fluvial landforms.

THEORIES OF SLOPE EVOLUTION

Considerations of the relationship between slope angle and slope stability inevitably lead to questions of the way in which slope forms change with time. Such changes are so slow that observational techniques are not able to provide direct evidence of the pattern of slope evolution. Several experiments in slope evolution have been carried out, Wurm (1935), making physical model experiments of the evolution of hillslopes by surface runoff, but these experiments could not reproduce the effects of vegetation or the influence of stream erosion at the base of the slope or along the valley axis. As an alternative to physical models, mathematical models of slope evolution have been widely used (Bakker and Le Heux, 1946, 1947; Van Dijk and Le Heux, 1952; Bakker and Strahler, 1956; Looman, 1956; Scheidegger, 1961; Gossman, 1970; Kirkby, 1971). The models developed by Bakker, Scheidegger and others seem to confirm Melton's observations on the relationship of slope evolution to the erosional environment. Four theoretical cases of slope development are generally discussed, the gradual reduction of slopes to successively gentler angles, the parallel recession of slopes, the development of equilibrium conditions under which some slopes will waste away while others are being created, and fourthly, the case of unequal activity, according to which each valley develops in relation to lateral river action. The last case stresses that a slope can recede only if there is a river (or waves) cutting away laterally at the base of the slope, while slope wastage without lateral action of a river (or waves) produces a slow decline of slope angle without a recession at the foot (Crickmay, 1960). Some theoretical models derived by Gossman (1970) are illustrated in Figure 28.

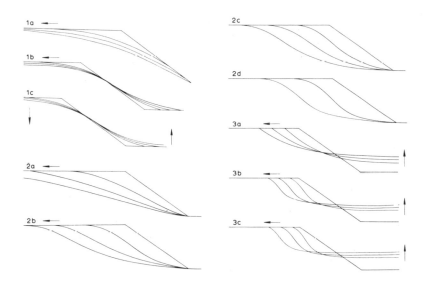

28   *Theoretical models of slope evolution proposed by Gossman*

The theoretical mathematical models tend to consider single processes as operative on slopes, yet all slopes are affected by a series of different processes, some operating more or less continuously while others are infrequent, episodic yet crucial for slope form, such as the landslips and slumps produced by mass failure. Recent efforts to produce hillslope process-response models (Carson, 1969; Kirkby, 1971; Thornes, 1971) suggest that the links between form and process may be developed into satisfactory pre-

dictive models. Both Carson and Kirkby indicate that slopes would
assume a curvilinear graded form which would be related to the
dominant process. They envisage a decline of slope under one set of
processes to a graded condition, which corresponds to various
threshold angles. A graded slope is one possessing a continuous
regolith cover, without rock outcrops, but the notion that there is
an equilibrium in the graded slope between rate of weathering and
rate of removal is said to be untestable. However, this is essentially
implicit in concepts of slope evolution, which all suggest that the
achievement of a balance between the factors of the slope system is
going to result in particular slope forms and particular patterns of
slope evolution.

The real validity of any theory of slope evolution depends on its
applicability to regional instances. In most morphoclimatic environments, evidence of slope evolution of each type may be found,
just as differing combinations of slope form are found. In the humid
tropics, Nossin (1964) and Swan (1970b) have found that in West
Malaysia slopes receded through the migration of the weathering
front back into the hillslope and the removal of the regolith by both
quasi-continuous processes of surface and subsurface wash and
infrequent landslipping. Dury (1959) cites instances of parallel
retreat of slopes in Britain, while Carson and others have produced
evidence of down-wearing of slopes. Evidently, much more is
involved in the processes of slope evolution than is apparent in the
general theories. The consideration of slopes in terms of process-
related models, such as the 9-unit landsurface model, even though
it involves sectionalisation of a continuous surface, may help to
develop explanations of slope forms in terms of the variables of
the denudation system and so facilitate the integration of slope
forms and valley forms into a general landscape morphology.

VI

CHANNEL INITIATION, HYDRAULICS AND SEDIMENT TRANSPORT

A fundamental change in the nature of processes moulding the landscape occurs at the point where ill-defined paths of water movement (unconcentrated flow) develop into lines of concentrated flow along defined channels. Unconcentrated flow on interfluves (unit 1 of the 9-unit model) is negligible, water moving vertically down the soil profile. Valley-side slopes show abundant signs of surface runoff with hollows, if not actual channels, to which such runoff converges. This contrast between interfluves and valley-sides is clearly apparent where a plateau ends abruptly in an escarpment as at Point Lookout in northern New South Wales or on the gritstone edges of the Pennines in England.

Controls of drainage density

In rugged terrain, interfluve slopes are commonly lacking, valley-side slopes intersecting in sharp ridges, such as those of the Bellingen and Macleay drainage basins east of the New England escarpment in New South Wales, producing what Cotton (1958) has termed fine-textured or feral relief (Plate 8). In the fine-textured relief of the North Island of New Zealand, the drainage density, that is the length of channels per unit area, ranges from 20 to 100 km per km^2, much higher than those of the humid south-west of Britain which range from less than 1·0 to nearly 10·0 (Gregory and Walling, 1968). Drainage density is a measure of the relationship between channel length and lengths of hillslopes between channels. This relationship between the efficient, concentrated flow of water in channels and the impeded, much slower movement of water on slopes, is fundamental for the rate and nature of landform evolution.

Drainage density is affected by both climate and lithology, especially permeability, and may be changed by an alteration to any of the factors which affect the runoff component of the hydrological cycle: precipitation, temperature or land use (Gregory, 1971). Drainage density is in part controlled by the number of points in the catchment where water becomes sufficiently concentrated to create a channel. The factors that affect the location, number and

8   *Finely dissected relief with* Kerbtal *in rain forest, Papua New Guinea —
a good example of feral relief (CSIRO photo)*

spacing of such points of runoff concentration are therefore fundamental to the pattern of landform evolution in a humid area.

Stream channel heads take varied forms. On the almost level
upland surfaces of parts of the Eastern Highlands of Australia,
water from swampy seepage zones feeds streams, as at the base of
granitic outcrops on the New England plateau near Round
Mountain. The precise point at which unconcentrated, diffuse flow
through the swamp grasses gives way to concentrated, linear channel
flow can be determined only by close field inspection.

In other situations, streams are fed by springs, sometimes by
large springs as in karst areas (Jennings, 1971) and basaltic
lava country (Ollier, 1969b), but more commonly by the outflow
from coalescing lines of subsurface water movement at a stream-
head hollow (Plate 9), such as those frequently found in deep regolith
material in the humid tropics, and such humid extratropical areas
as the coastal lands of New South Wales and the North Island of
New Zealand. On steeper slopes, streams may emerge from beneath
a mass of rock debris occupying the highest parts of a first order
valley. Such chokes of core-boulders are common in forested
granitic terrain throughout the humid tropics and in eastern
Australia. In temperate environments, such as the New England
area of New South Wales, the boulder choke in the head of the
stream gives way down-valley to fill of granitic colluvium through

9 *Stream head hollow controlled by subsurface water movement on recently cleared land in Johor, Malaysia (Photo by S. B. St C. Swan)*

which water may move along the axis of the valley in subsurface pipes. As the valley gradient decreases, subsurface pipes either collapse or are cut by a gully headwall and open channel flow begins.

Returning to a stream head after a dry period, the field worker will often find that the stream no longer flows from so high up in its valley. Yet, under exceptional rain conditions, concentrated flow may actually begin some distance upslope of the topographic valley or gully. Streams in the Paris Basin vary in length according to season, channel flow often being discontinuous, water rising in sections of the channel, sinking some distance down-valley and reappearing once again further downstream (Cailleux, 1948). Such phenomena may be due to subsurface flow and the occurrence of freeze and thaw cycles providing meltwater in winter.

Variations in the length of stream flow within a channel may be due to other factors. In a West Malaysian gully system cut in thin bands of deeply weathered shales and sandstones, less than 1 mm of rain in the previous two hours produced flow just above a small nick point, but between 14 and 20 mm in the previous two hours were needed to produce runoff from the channel heads approximately 35 m above the nick point (Morgan, 1972). Extremely intense rainfalls produced organised lines of flow above the actual gully heads. Rain sufficient to cause erosion of the gully sides would occur approximately once every 60 days.

EVOLUTION OF CHANNEL HEADS

The variation in length of channel flow is a response to variation in the input of water to the surface and subsurface runoff components of the basin hydrological cycle. In the Paris Basin example input from snow melt was a major factor, while the more rapid variations in West Malaysia are in response to stormwater inputs. Channel head position is determined by the character, magnitude and frequency of runoff, while channel head form varies within major climatic zones as a result of structural influences and the character of the weathering environment. Although in some cases the position of the channel head is determined solely by the character of overland flow after heavy storms, the pattern of subsurface water movement is a more important control in most humid environments. Channel initiation from overland flow was described by Horton (1945) and is the simpler case. The work on subsurface flow must be regarded as an extension and complication of Horton's overland flow model.

The Horton model of channel initiation

If water is poured on to a bare, unconsolidated, sloping surface, rills will develop on the slope as soon as the flow becomes sufficiently turbulent to entrain debris. However before water will begin to flow over the surface of a slope, rainfall must have been sufficient to meet the infiltration capacity of the soil or to have fallen at a rate exceeding the infiltration rate. Near the crest of the slope the depth of runoff water will always be shallow, but as the slope lengthens, and declivity probably also steepens, and the depth turbulence of surface runoff will increase until, at a certain critical distance down-slope, erosion will begin and rill channels develop (Horton, 1945). Thus, where overland flow is dominant, a belt of no erosion at hill crests may be recognised (Fig. 29). Such zones are often clearly apparent where recently deposited volcanic ash has been dissected by running water before vegetation has become established.

Surface runoff starts at the slope crest as true sheet flow, without channels. However, as no natural slope is completely smooth, accidental concentrations of sheet flow occur. The rill channels which start to form as a result of these points of increased water movement are usually relatively uniform, closely spaced and nearly parallel channels of small dimensions which are initially developed by sheet erosion on a uniform, sloping, homogeneous surface (Plate

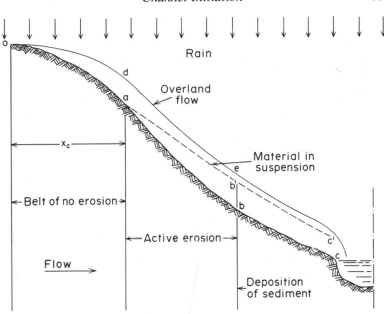

29   *Horton's diagram to illustrate overland flow and the initiations of erosion on hillslopes*

10). Once a rill has formed sheet flow coming down the slope upstream from the head of the rill will be deflected towards and diverted into the rill channel, providing a means for the erosion up slope of the head of the rill to the point where water running off after the heaviest rains is just able to create sufficient turbulence to maintain a defined channel. Such a position is a function not only of rainfall quantity and intensity, but also of the angle of slope, the nature of the soil materials, the vegetation cover and the roughness of the slope surface.

Debris transport by overland flow and in rills depends on the depth of flow, degree of turbulence and supply of fine particles by such processes as raindrop splash. As discussed earlier, much of the material eroded and transported by surface runoff is redeposited further downslope. In the bare soil situation, the gullies or rills developed on the steepest section of a slope become less incised and die out as siltation of debris chokes the channel and water spreads over and infiltrates into alluvial material at the foot of the slope (Plate 11).

LEWIS AND CLARK COLLEGE LIBRARY
PORTLAND, OREGON 97219

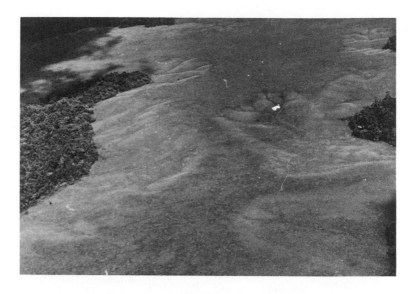

10 *Overland flow and drainage initiation on the Sepik Plain between Sepik and Adelbert Rivers, Papua New Guinea (CSIRO photo)*

11 *Rills on bare soil between row crops on the Atherton Tableland, north Queensland and redeposition of material at the base of the slope*

In intermediate situations, where the soil is not entirely bare, and where there is not sufficient downslope decrease in gradient, the rills and gullies develop into an integrated channel network. In this case, the silt-laden surface storm runoff pours into rills and gullies, creating such turbulence on reaching the gully floor that the head wall and sides of the gully are undercut, with consequent gully lengthening and widening. The debris from surface runoff and gully enlargement are together carried into higher order channels to be deposited further down the stream network or carried out of the drainage basin altogether.

As described in Chapter IV, under dense vegetation cover the removal of debris by surface runoff is limited by the plants. Rill development is minimal, severe soil loss only occurring when the vegetation is disturbed, through tree fall or through disruption of the grass soil cover and stripping of turf from the slopes (Plate 12).

12 *The extreme result of severe soil erosion following vegetation disturbance along the old coach road from Mareeba to Thornborough, north Queensland. Remnants of the original soil surface appear as isolated buttes, while dead trees, uprooted by the headward extension of gullies, litter the landscape.*

The throughflow model of stream initiation

Subsurface water movement may take several paths, with many

areas having both surface and subsurface runoff, as in the Plynlimmon catchment of central Wales where seven principal flow routes have been identified:

a) laminar surface runoff;
b) ephemeral surface channels;
c) subsurface laminar flow on the interface of a B horizon pan and a peaty A horizon;
d) subsurface pipe flow on the same horizon;
e) laminar flow on the interface between a boulder clay (glacial till) and its thin mantle of colluvial material; this lies beneath a clayey 'B' horizon (with shrinkage cracks) and a peat cover;
f) pipe flow on the same horizon;
g) laminar flow in the surface zone of the bedrock.

In an impermeable catchment, with a well-developed soil cover which originally supported woodland on the Mendip Hills in southwest England, Weyman (1970) found throughflow to be the only process of water transmission effectively contributing to streamflow (Fig. 9). The 'quick response' of stream discharge to rainfall in the Mendip catchment is probably the result of rapid increases in throughflow output as saturated conditions extend upslope. In a series of geologically contrasted catchments in south-east Devon, in the same part of Britain, throughflow and saturated overland flow were found to be the main controls of storm runoff and erosion (Walling, 1971).

Gully development by the collapse of subsurface pipes, as described in Chapter IV, is one mechanism of drainage density control by subsurface water movement. Another is the emergence of water from the back wall of hollows where subsurface water has converged and so saturated the ground that seepage occurs at the surface. Alternatively, a subsurface pipe may bring water into the hollow. The stream head is extended at time of high runoff, when the back wall is eroded by discharge from the pipe and collapse of adjacent material, but the hollow tends to be filled in during drier periods when runoff is insufficient to evacuate material falling into the hollow. The position of the stream head is thus an adjustment between the tendency of channel extension to form a hollow of increasing concavity, the tendency of mass movements on the surrounding slopes to fill such a hollow (Kirkby and Chorley, 1967) and the trend of recent climatic events.

The partial area model of storm runoff

Both the Horton overland flow and the throughflow models of storm runoff and channel initiation suggest that storm runoff may originate in a widespread fashion everywhere that storm rains occur within a drainage basin. However, storm runoff may be generated only in small, but variable, areas of catchment. As throughflow and surface runoff converge in natural declivities, some parts of a catchment area will become so wet that only a relatively small input of extra moisture into the soil will produce a large relaxation of moisture tension in the soil pores and so cause a rapid rise in the level of the water table. This rapid rise may accelerate runoff to stream channels. After exceptional storms, the rise in the saturated upper, perhaps perched, water table of the throughflow zone extends along the channel network into the first order valleys in which channels may not exist, producing large quantities of surface runoff. During subsequent days, the saturated zone decreases in area until only the perennial channels are supplied with water. This partial area model emphasises the changing nature of sources of streamflow and the importance of the saturated areas which develop near the perennial channels during the storms as sources of storm runoff (Hewlett and Nutter, 1970).

The partial area model also recognises the expansion and contraction of the functioning drainage net, the actual change in drainage density from one storm to another (Fig. 30). Thus the drainage net is infrequently used to maximum capacity. Rare, extremely heavy storms, or exceptionally rapid snow melt will produce storm runoff which extends concentrated flow beyond the headward limits of the channel network. The stream heads thus are regulated by present-day hydrological conditions. In areas where greater runoff prevailed in earlier times, extensive dry valley networks may exist (Gregory, 1971) as topographic evidence of climate change. The changes in runoff conditions following removal of vegetation may lead to an extension of the channel network, as storm runoff volumes increase and water discharges from slopes more rapidly (Fig. 31).

The dynamic nature of stream channel heads

The hydrological control of stream head positions shows that stream channels are the product of conditions on the slope above them. As Arnett (1971) found in the Rocksberg Basin in south-east

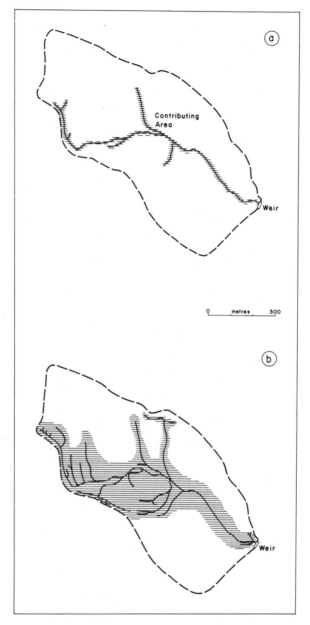

30　*Variation in the extent of the area contributing runoff after different storms.*
　　(After Gregory and Walling, 1968).

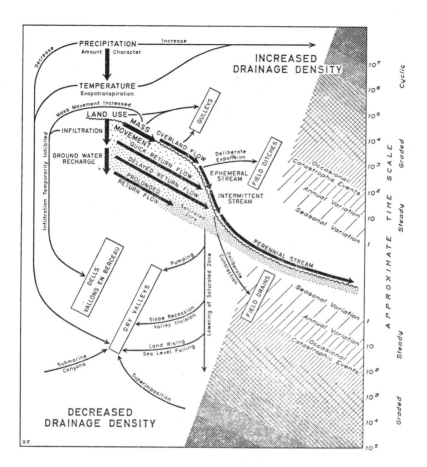

31 *A model of explanations suggested to account for drainage density changes.
The various explanations are positioned on the diagram with respect to
the geomorphological time scale represented by shading. The types of
water flow in the basin, represented in the centre of the diagram, are
based upon D. G. Jamieson and C. R. Amerman 'Quick return subsurface flow', Journal of Hydrology, 8 (1969), 122–36. Reproduced
with permission from Gregory (1971).*

Queensland, the form of the slope and nature of the slope materials
control the channel initiating process. However, there is a feedback
in the system, or a process-response mechanism, by which the extension of the stream head hollow may cause gravitational mass movements on the slope above the hollow which gradually tend to refill

the hollow itself. Stream head hollows and other forms of channel head are thus the topographic expression of the dynamics of a drainage basin process-response system which links slope and channel processes (Chorley and Kennedy, 1971). Obviously, such controls of the heads of stream channels only permit the headward extension of streams under special conditions of storm runoff input. Notions of streams cutting back to 'capture' other drainage lines prevalent in some older texts should therefore be critically examined.

SOME FUNDAMENTALS OF HYDRAULICS

The flow of water across the land surface to streams and along the open channels of streams and rivers, and the associated transport of solid particles and dissolved matter, is governed by a series of stresses which are essentially part of the interaction between *gravitational forces* pulling the water downslope, *frictional forces* retarding the movement of the water and the *volume of water* available for discharge. In hydraulic engineering terms the forces acting on water flowing in an open channel are: surface tension; the component of the water's weight acting in the direction of the bed slope; shear stresses developed at the solid and free surface boundaries; internal inertial forces due to the turbulent nature of the flow; normal pressure at the walls and bed, particularly in areas where channel dimensions are liable to change; and the movement of sediment (Sellin, 1969). The mutual interaction of these factors accounts for the complexity of water movement in rivers. To unravel this complexity the factors have to be discussed separately and then be brought together successively. However, as river channels are the product of so many interacting factors, the relationships and constants discussed below are the empirical results of many measurements, not precise physical quantities.

The actions of frictional forces along the banks and bed of the channel and at the air-water interface and of internal inertial forces cause an uneven distribution of water velocity in any given cross-section. If velocity is measured at any given vertical, the velocity at the bed is minimal, but increases upwards to a maximum just below the surface. In a cross-section through a straight reach of a channel velocity is normally greatest just below the surface in the centre of the stream. Not all streams have the typical velocity distribution at

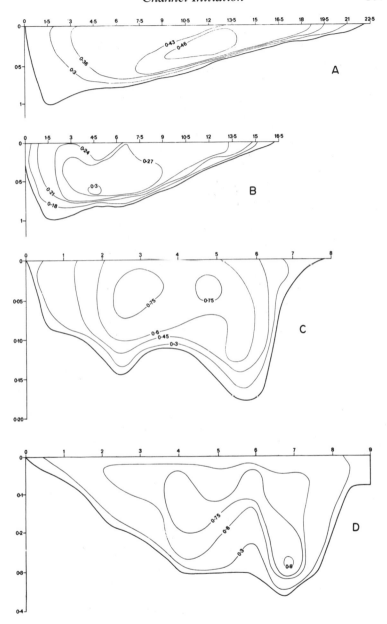

32 *The distribution of velocity at various discharges in Behana Creek at Aloomba, north Queensland. Horizontal axis is width in metres, vertical axis depth in metres.*

all stages of flow (Fig. 32). The actual distribution depends on the configuration of the channel bed and the shape of the channel upstream of the velocity measurement site.

Flow in channels with an immovable bottom

The pattern of velocity distribution in a stream channel shows clearly that velocity increases where frictional forces are least. Engineers have long tried to account for this by using a formula to calculate velocity taking into account the cross-sectional area of channel flow, the slope of the channel, and the wetted perimeter. The formula used is the Chézy formula

$$V = C\sqrt{RS} \tag{1}$$

where V = velocity
C = constant (Chézy's C)
R = hydraulic radius $\underline{A}$ (the cross-sectional area of the
P_*
channel flow (A) divided by the length of the wetted perimeter in a cross-section (P_*)
S = $\sin \theta$, θ being the angle to the horizontal at which the channel slopes (Fig. 33).

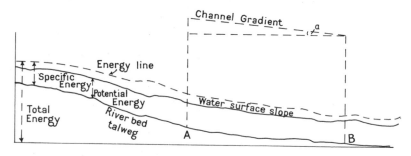

33 *Channel gradient, water surface slope and specific energy. The channel gradient between A and B is measured as the angle A.*

Chézy's C is derived from a formula:

$$C = 1 \cdot 5 \frac{R^{\frac{1}{6}}}{n} \tag{2}$$

in which C is recognised to be a function of the hydraulic radius and a roughness coefficient n. By substituting in equation (1), the widely used Manning formula is derived:

$$V = \frac{1 \cdot 5}{n} R^{\frac{2}{3}} S^{\frac{1}{2}} \tag{3}$$

Determination of the roughness coefficient n is not easy and, at the present state of knowledge, the selection of roughness coefficients for natural channels is an art, in which accuracy is acquired with experience (Barnes, 1967). The coefficient expresses the condition of the bed and banks of a river with regard to the energy losses caused by irregularities and obstructions on the channel perimeter. Table 8 sets out the range of values of n for given channel bed types and conditions, but a much better guide to Manning's n is provided by the U.S. Geological Survey's set of pictures and diagrams of stream channels for which n has been determined by using an energy equation (Barnes, 1967).

TABLE 8 **Roughness coefficient, n, for the Manning formula**

Type of description of channel	n values
Channels, lined	
Glass, plastic, machined metal	0.009–0.013
Concrete	0.012–0.018
Metal, corrugated	0.921–0.026
Wood	0.011–0.015
Channels, vegetated	
Dense, uniform stands of green vegetation more than 25 cm long	
Bermuda grass	0.04–0.20
Kudzu	0.07–0.23
Lespedeza, common	0.047–0.095
Earth channels and natural streams	
Clean, straight bank, full stage	0.025–0.040
Winding, some pools and shoals, clean	0.035–0.055
Winding, some weeds and stones	0.033–0.045
Straight, some rocks and/or brushwood	0.050–0.080
Sluggish, river reaches, weedy or with very deep pools	0.050–0.150
Straight, very rocky or with standing timber	0.075–0.150
Pipe	
Asbestos cement	0.009
Cast iron	0.011–0.015
Floodplains	
Short grass pasture	0.025–0.035
Mature crops	0.025–0.045
Brushwood	0.035–0.070
Heavy timber or other obstacles	0.050–0.160

After Sellin (1969) and Schwab, Frevert, Barnes and Edminster (1971).

An alternative to the estimation of Manning's n is the examination of the relative roughness of a channel, the ratio of depth to the size of the roughness elements. As most rivers contain varied sizes of material some expression of particle size has to be used to provide the magnitude of the roughness elements. During investigation of Brandywine Creek, Pennsylvania (Wolman, 1955), the particle diameter (D_{84}) equal to or larger than 84 per cent of the particles was used. This arbitary measure, one standard deviation larger than the mean size, assuming a normal distribution, provides a single bed particle size parameter, recognising that particles larger than the median size play a major role in creating resistance to flow.

The Manning equation is a good approximation to the more refined relationships between the Darcy-Weisbach resistance coefficient, Reynolds number, and the relative roughness (Kolosius, 1971). The Darcy-Weisbach resistance coefficient, a dimensionless parameter, f, used to express resistance to flow in conduits or open channels and proportional to gdS/V^2, is obtained from the formula:

$$\frac{1}{f} = 2 \log \frac{d}{D_{84}} + 1 \cdot 0 \qquad (4)$$

where g = the acceleration due to gravity
d = the depth of flow
D_{84} and f are as defined above and S and V are as defined for equation (1).

In contrast to the Darcy-Weisbach coefficient, the Reynolds number estimates resistance from the velocity and hydraulic radius

$$Re = \frac{VR}{r} \qquad (5)$$

where v = kinematic viscosity of the fluid
Re = Reynolds number
V and R are as defined in equation (1).

Manning's n, the Reynolds number and the Darcy-Weisbach resistance coefficient are interrelated, but the nature of the relationship depends on the character of the channel bed and shape.

Turbulent flow in channels with movable beds

Relationships developed for channels with an immovable bottom are not readily applicable to natural channels as interaction occurs between the bed and the flow. Although for natural channels resistance is probably most closely approximated by the Darcy-Weisbach coefficient, in practice the more easily determined Manning's n is more frequently used. Even so, errors in the value of n of the order of 10 per cent arise from such factors as the growth of aquatic plants. In streams with coarse material on their beds, such as Brandywine Creek, conditions approximate to those of uniformly distributed roughness characteristics and total flow resistance is a function of relative roughness. In streams with fine material beds, however, channel bed configuration changes as flow varies, affecting the value of n, which is not the same for all seasons and conditions of flow.

The changes in energy distribution in a cross-section as depth or width vary may be discussed in terms of specific energy. Specific energy is total head above the floor of the channel, head being the hydrostatic pressure of a liquid divided by the specific weight of the liquid. Specific energy is calculated as:

$$E = \frac{Q^2}{2gA^2} + d \qquad (6)$$

where E = specific energy
 Q = total discharge
 A = cross-sectional area of flow
 g and d are as defined in equation (4)

Where contractions occur in open channels, velocity of the flowing water increases, but the water surface is lowered. The same discharge passes through a smaller cross-sectional area at greater velocity. Thus two different depths can exist at a given specific energy and discharge. The depth is small when the velocity is great, but is great when the velocity is small.

This variation of energy with depth is indicated by the specific energy curve in Figure 34. The point on the curve where the specific energy is least is termed the *critical depth*. Flow at depths greater than the critical is termed *subcritical*, while at depths less than the critical, the flow is termed *supercritical*. The relationships between velocity and depth flow in Figure 35 show that there are four regions of flow in open channels: subcritical laminar, supercritical laminar, subcritical turbulent, and supercritical turbulent (Allen, 1970). The turbulent regions are those affecting stream channels. Subcritical turbulent flow is termed *lower flow regime*, while supercritical turbulent flow is termed *upper flow regime*. (Table 10).

Humid Landforms

TABLE 9 **Variation of total sediment concentration with flow regime,
bed roughness and particle size**
(based on laboratory data by Simons and Richardson, 1962)

Regime	Forms of bed roughness	Total sediment concentration mg/l	
		0·28 mm sand	*0·45 mm sand*
Lower flow	Plane	0	0
	Ripples	1–150	1–100
	Dunes	150–800	100–1200
Transition		1000–2400	1400–4000
Upper flow	Plane	1500–3100	—
	Standing wave	—	4000–7000
	Antidunes	5000–42 000	6000–15 000

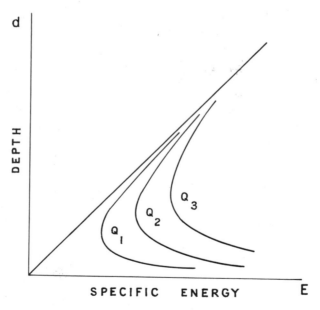

34   *The specific energy curve.* Q_1, Q_2 *and* Q_3 *represent different discharges*

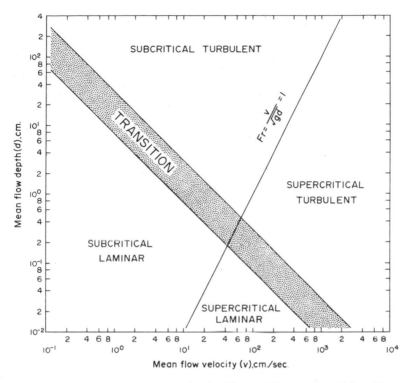

35 *Relationship between velocity, depth of flow and flow regime. (After Allen, 1970).*

SEDIMENT TRANSPORT IN STREAM CHANNELS

Individual particles in a movable bed channel are subject to three kinds of forces, *gravity* forces affecting the settling velocities of particles in water; the *scouring* effect acting on particles sitting at the bottom of a stream bed; and the *lifting* force affecting the transportation of sedimentary particles in suspension. As soon as water starts to flow over particles in a river bed, hydrodynamic forces act on the particles in contact with the water. An increase in flow intensity causes an increase in the magnitude of these forces. Thus at any given cross-section in a particular stream, a condition is eventually reached at which the particles in the bed are unable to resist the hydrodynamic forces and are thus dislodged and set in motion. Not all particles are moved at once, the critical force required to initiate motion varying with grain size.

If the particles set in motion when the *critical condition* of the bed is exceeded roll, slide or sometimes jump (saltate) close to the channel bed, they are said to be transported as *bed load*. If, however, the entire motion of the solid particles is such that they are surrounded by water, they are said to be carried as *suspended load*. Bed load is moved by scour effects through the stream boundary, stress acting on the channel perimeter and forcing particles upwards as they collide and slide against each other. Gravitational forces pull the particles down the channel, down the banks and back to the bed. Other particles are set in motion by collision and pushing. Suspended load is kept in motion in the water body by turbulent eddies in the fluid which impel fine particles upward. Transport in suspension is the dominant mode of carriage of quartz-density solids whose diameter is less than about 0·5 mm (Allen, 1970).

Physically, bed load may be defined as that part of the load whose normal immersed weight component is in normal equilibrium with the dispersive stress between sheared grains. This stress is transmitted downwards via the dispersed grains to the stationary grains of the bed upon which it therefore ultimately rests (Bagnold, 1956). Suspended load is that part of the load whose immersed weight component is in equilibrium with a normal fluid stress originating in impulses by turbulent eddies.

For some geomorphological purposes, the distribution between bed load and wash load is more appropriate than the definitions given above. Bed load is the sediment transported in a stream by rolling, sliding, or skipping along the bed (Colby, 1963), whereas wash load, or fine material load, consists of sediment so fine that it is about uniformly distributed in the vertical and is only an inappreciable fraction of the total sediment resting on the stream bed. The size of particles in wash load at a given time in a given cross-section is a function of stream flow and of sediment supply. In most streams, wash load is limited by its availability in the catchment area (Einstein, 1964).

Saltation as a separate mode of sediment transport in streams is unimportant (Graf, 1971) but its effect is included in bed load motion. Although saltation during wind erosion is well established (Bagnold, 1941), there is some doubt whether the apparently similar processes observed in water are saltation in the strict sense of the word, as the relatively high density of water has a strong damping effect on the impact and momentum transfer between sand grains (Sellin, 1969).

The finest particles carried by streams are in colloidal suspension. Such particles may range in size from nearly that of ions in true solution (around 10^{-1} nanometres in diameter) up to about 2×10^2 nanometres (Hem, 1959). As the finest filters used to separate suspended particles from stream water samples cannot hold back material finer than 10 nanometres, much colloidal matter is measured as dissolved solids and no exact boundary between the sizes of particles considered to be in solution and those in colloidal suspension can be given.

Dissolved matter accounts for the major part of the total load carried by many rivers. Ions in solution are carried by the stream as rapidly as water flows along the channel, but are constantly modified in character by interaction with organic matter and solid particles, together with physical changes due to evaporation or additional water input into the channel. Quantities of solutes are controlled by the lithological and climatic characteristics of the catchment area.

Rates of sediment transport related to discharge

The quantity of sediment being carried by a stream at any one time is a function of sediment supply rather than the transporting power of the stream, save in the case of bed load. Suspended sediment loads and dissolved solids vary from season to season, and storm to storm. Depending on rainfall distribution within a catchment, tributaries carry differing concentrations of sediment at similar discharges. Figure 36 shows how solute concentrations drop slightly as discharge increases, while suspended sediment concentrations rise rapidly, but reach a peak before the maximum discharge. The early peak of suspended sediment concentration shows that sediment is supplied at a rate well below the transporting capacity of the river, thus confirming Einstein's suggestion (1964) that the wash load of a stream is limited by the availability of sediment in the watershed. Only in sand bed streams, where the sediment supply from bed and banks is abundant, does a reasonably close relationship between stream velocity and the transport of wash and bed load material exist. Such relationships defined for one cross-section may apply for other sand bed streams, providing bed material sizes and water depths are similar and the wash load is not unusually high (Colby, 1964).

At any given station on a river, the suspended load (the suspended sediment concentration multiplied by the discharge) increases as a power of the discharge. On the Brandywine Creek at Wilmington,

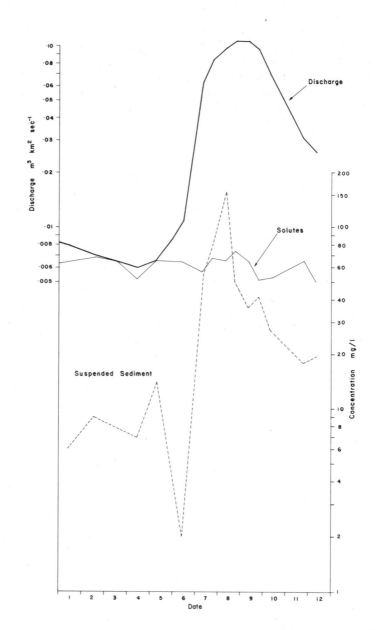

36 *Discharge, solute and suspended sediment concentration during the passage of cyclone Flora, Barron River, north Queensland, December 1964*

Delaware, U.S.A., the suspended load increases as the 2·37 power of the discharge. This is the value of j in the empirical equation:

$$Q_s = pQ^j$$

where Q_s is the suspended load, Q is the discharge, and p is a constant (Wolman, 1955). Similar values for some Malaysian rivers are given in Table 10. The overall linear logarithmic relationship between load and discharge does not reveal the variation of load during the rising and falling stages of stormwater discharge and variations in the supply of material to rivers from storm to storm and season to season. In temperate climates differing load : discharge relationships can be established for different times of the year (Imeson, 1970; Walling 1974), while in a sequence of storms the suspended sediment load at a given discharge decreases from a high level in the first storm as material for transport becomes less readily available in the catchment (Negev, 1969).

In contrast with the role of rainfall and catchment characteristics in suspended sediment transport by streams, bed load transport is more closely related to channel hydraulics. While a direct relationship between velocity and bed material transport may exist in sand bed streams, coarser bed material moves in an irregular manner. In ephemeral streams in New Mexico the average distance of movement of particles weighing 200 to 1300 gm is broadly related to discharge, but a greater flow velocity is required to shift particles of a given size which are close to one another than if they are spaced far apart (Leopold, Emmett and Myrick, 1966). In Cache Creek, west of Sacramento, California, bed load transport increases as discharge increases, but falls from about 85 per cent of the total sediment discharge at 3·1 m³ sec⁻¹ to less than 3 per cent at 283 m³ sec⁻¹ (Lustig and Busch, 1967).

Movement of pebbles and gravel by floodwaters in mountain streams in Japan is only partly affected by flood discharges and stream flow velocities (Takayama, 1965). Only in the two largest of twenty flood flow studies were a large proportion of the pebbles of 20 to 150 mm diameter moved as much as 100 mm from their starting point. The movement of individual pebbles is an episodic, short-distance process. While a 5 m³ boulder was moved by a 3 m deep flood with a peak velocity of about 5 m³ sec⁻¹ (Wolman and Miller, 1960), the distance any such particle will be carried is small. Less than 1 per cent of the violet coloured volcanic rocks which were

TABLE 10 **Values of the exponent j in the suspended sediment load-discharge relationship for selected streams in the Sungai Kelang catchment, Malaysia**

Stream	Catchment area km²	Exponent	Comments
Gombak at 13th milestone	41.3	1.29	Jungle headwater stream with little human interference
Pasir at 13th milestone	3.3	2.81	Small tributary draining large road cutting supplying abundant sediment
Anak Ayer Batu at Jalan Damansara	3.3	2.54	Small tributary draining land cleared for heavy development with high sediment concentrations
Lake tributary at Jalan Damansara	0.6	2.32	Drains new housing area with abundant sediment sources
Gombak at Jalan Pekililing	140.0	1.00	Affected by urban runoff and erratic sediment laden discharge from alluvial tin mines at high and low flows
Batu at Jalan Pekililing	151.2	1.12	Also affected by tin mining and urban development
Kuyan at Jalan Puchong	82.0	0.92	Affected by discharge from tin mines
Kelang at Puchong	1196.6	1.18	Main stream just above tidal limit after receiving all tributary runoff from sources indicated above

The variety of values of the exponent j as a result of differences in sediment supply in this one humid tropical river basin should be contrasted with general statements on values of j such as that suggesting a value between 2 and 3 for rivers in the western U.S.A. (Graf, 1971).

swept down the Guil River in the French Alps by the catastrophic floods of 1957 were carried more than 6 km from their source (Tricart and Vogt, 1967).

Movement of coarse bed load takes place by load substitution. Individual particles are lifted and rolled a short distance before being dropped to replace other particles which have been taken up further downstream. Thus the composition of coarse bed material

13 *Rounded granite corestone boulders in the Sungai Gombak at 12½ mile-
stone near Kuala Lumpur, Malaysia. These corestones have rolled into
the channel from the valley-sides and are seldom, if ever, moved by flood
flows. Their chemical and mechanical disintegration yields quartz grains
to the stream.*

changes but slowly in a downstream direction. Pebbles do not
change shape rapidly, but there is a gradual decrease in size downstream, the less resistant pebbles being worn away most rapidly and
eventually eliminated through abrasion and chemical decomposition. Thus if further pebbles are not supplied by tributaries or from
the banks, the most resistant pebbles, usually those of quartzite
and quartz, become dominant (Plate 13).

Modes of bed load movement also vary according to catchment
characteristics and the hydrological regime of rivers, as shown by
the comparison of the Rivers Rhine and Durance in western Europe
(van Andel, 1951a, b). Reworked old valley deposits in the Rhine
between Strasbourg and Wageningen are subject to local erosion
and redeposition, particles being transported downstream extremely
slowly. On the other hand, transport of material by the Durance
is essentially a continuous mass effect, only interrupted during the

relatively short periods of low water. At each flood all sediments in the Durance are thoroughly mixed and rapidly transported in a poorly sorted mass.

Tricart (1961) emphasises that while the catastrophic transport of enormous amounts of coarse debris (*charriage en vrac concentré*) profoundly changes the nature of river channels, it is of relatively small significance in the long term transport of material, being limited in time and in distance of movement. In terms of tons of load transported per kilometre the amount of movement is quite small. Pardé (1954, 1958), however, argues that in mountainous streams movement of pebbles may account for as much as 200 m^3 km^2 yr^{-1} of removal from the catchment, and that bed load transport would equal, or exceed by several times, the transport in suspension. He quotes examples of floods in mountain torrents in Japan and California in which enormous volumes of boulder size debris are transported. Nevertheless, the long term significance of the movement of this coarse material, compared with transport of suspended material and dissolved matter, remains uncertain.

Particle shape may affect the movement of sediment as flat particles tend to become imbricated. Elongated pebbles are usually arranged with their longest dimension at right angles to the direction of flows and with a downward inclination upstream, and when the pebbles are rolled they tend to move with their long axes normal to the transporting force. The pebbles making up the beds of New Mexico irrigation channels (Lane and Carlson, 1954) usually rest on sand, which fits closely around the lower portion of the pebble. The presence of this sand means that a small scale, short duration turbulence in the streamflow may not be able to lift the pebble into transport, whereas a large scale turbulence acting for a long time may be able to shift it. Such an arrangement of bed material exists in many rivers of eastern Australia, where metamorphic rocks yield flat-shaped particles in streams, or in the highlands of the south-east, where frost action during past periods of periglacial climate has split pebbles in flat fragments. The beds of such streams have in fact a form of paving and, as Kajetenowicz (1958) emphasises, significant bed load transport does not occur until this paving is disrupted. Such a disruption occurs only when the water penetrates through the sand beneath the pebbles with enough force, or sufficient dispersive stress, to lift the pebble up against the turbulence of the water above and the other stresses which are tending to retain the pebble in its imbricated position.

Once the disruption of the paving is begun, however, there may be widespread movement of material. The precise hydraulic and hydrologic conditions under which such movement would occur are almost impossible to predetermine, field observations through several flood flows being necessary to ascertain when such changes take place.

So far in this chapter the effects of river bed shape on flow velocity and the effects of flow velocity on sediment transport have been discussed. However, all these effects react mutually with each other and the actual shapes and patterns of river channels are the result of a complex set of adjustments between these effects. The adjustments between water discharge, velocity and channel shape were recognised by Leonardo da Vinci:

> If the water in the river neither increases or decreases, it flows past each point in its width with a constant mass and with varying speed or slowness arising from the changing narrows and broads on its course (*Codex Atlanticus*, 287 r-b) (Heydenreich, 1954).

Leonardo's comment that: 'The velocity of a stream decreases in proportion to the increase in its breadth and depth' (*Codex Atlanticus*, 80 r-b) foreshadows the hydraulic geometry relationships which show that at a given station on a river, a change in discharge produces almost instantaneous adjustment in water depth, width and velocity (Leopold and Maddock, 1953). However, rivers also adjust to changes in sediment load, to the character of the material through which they flow and to stream gradient. Stream channels cut in hard rock undergo little change over short time spans while channels cut in alluvium may undergo massive scour, fill, bank erosion and deposition during the rise and fall of a single flood. As the type of adjustment in alluvial channels varies with the cohesiveness of the bank material and the predominant mode of sediment transport (Schumm, 1963a), three types of alluvial channel have been suggested: *suspended load channels* transporting perhaps 0 to 3 per cent bed load, *bed load channels* transporting more than 11 per cent bed load, and *mixed load channels* transporting from 3 to 11 per cent bed load (Schumm, 1971).

Flow conditions and bed forms in alluvial channels

Streams with sandy beds tend to develop ridges or ripples in the sand beneath the flowing water. Sometimes the ripples look very much like the ripples on a beach, but ripple forms change or the flow velocity changes in a systematic manner.

In the lower flow regime (Fig. 35) flow is calm and undulations of the water surface do not correspond to undulations of the bed (Simons, 1969). Distinct bed forms develop under lower flow regime in alluvial channels when the diameter of the bed material is less than 0·7 mm. Ripples are common on stream and river beds (Plate 14), while in deep rivers dunes 10 to 20 mm in height and 100 or more metres in length have been observed (Simons, 1969). In rivers with ripples or dunebeds, the value of Manning's n may decrease as depth of flow increases when the bed is formed of fine sands, long dunes formed by fine sands having smaller n values than ripples. However, with larger sand sizes, dunes are shorter and more angular presenting higher n values.

In the transition flow zone bed forms depend on the way in which discharge is changing, with n values ranging from the largest value for the lower flow regime to the smallest value for the upper flow regime.

14 *Ripples on a medium to coarse sand bedded stream in low flow regime in Moory County, South Carolina (Photo by J. N. Jennings)*

In upper flow regime motions in the stream become associated with the movement of sand on the bed and the development of antidunes, sinusoidal bed and water surface waves of low amplitude being broadly in phase with each other. Such antidunes may be seen in shallow alluvial rivers or gullies with fine sand. While the antidunes in gullies rarely exceed 20 cm in wavelength those in rivers may reach 10 m in wavelength and up to 2 m in height (Allen, 1970). When antidunes reach a certain critical height they break, and the water surface is flat and smooth for a short while. Breaking antidunes and associated water waves dissipate large quantities of energy which is reflected by an increase in the value of Manning's n. The increase of n is proportional to the amount of antidune activity and the amount of the stream channel occupied by antidunes.

In channels containing bed material of widely varied sizes, the relationships between flow, bed form and roughness are more varied than in the simpler case of sand sized bed material. However, most streams containing coarse ill-sorted materials have a series of pools and ripples with associated changes in velocity and depth relationships.

When bed roughness consists of ripples and/or dunes under lower flow regime, the ability of a stream to transport bed material is relatively small (Table 9). Sediment transport under upper flow regime is much greater, the equivalent of up to 100,000 mgl^{-1} of bed and suspended material being moved under antidune conditions when sediment supply is abundant.

Flow conditions, sediment and bed forms in hard rock channels

Whereas changes in the beds and alluvial channels are largely the result of the impact on abundant easily moved sediment of stresses derived from the energy of flowing water, changes in rock-cut channels depend on the nature of loose particles carried by the stream, the character of the rock into which it is cut and the turbulence and velocity of the water. Water acts both chemically and mechanically on the rock over which it flows. Chemical reactions of the type described in Chapter III occur in the rock immediately below the stream, especially along joints and bedding planes. Where the latter have been weakened sufficiently, the scour forces which lift sedimentary particles off the bed may also lift weathered rock fragments. Rubble so detached will be carried downstream, exercising physical drag on the surface of the bed as it does so. The physical drag causes wear, or *abrasion*, on both the loose fragments and the bed itself (Plate 15). Such abrasion is part of the process of *corrasion* by which the channel floor is gradually lowered over long periods of time. When the particles are harder than the rock bed, possibly because they are derived from hard rock outcrops upstream, they may rub grooves into the rockbed, sand grains sometimes causing striations in hard rock channels (de Martonne, 1951).

The clearest evidence of downward corrasion in hard rock channels is the development of *swirlholes*, or potholes, round vertical shafts up to several metres across resulting from the whirling round of pebbles by a turbulent eddy in a crevice in the rock (Plate 16). Swirlholes often occur where resistant rock bars obstruct channels carrying coarse sand and gravels, as at the Nile cataracts and at rapids in humid tropical rivers such as those of the Sungai Dong in Malaysia. Swirlholes, however, are also common in steep channels carrying large volumes of water, as in north Taiwan, where swirlhole development is closely related to joints in the rock

15   *Abraded bed of Mount Pierce Creek in Mount Pierce Gorge, north Western Australia in the dry season. Note how the gravel has been sorted by eddy motion in the swirlhole (Photo by J. N. Jennings).*

(Tschang, 1957). Rounded pebbles, usually of resistant material such as quartzite, generally lie at the bottom of swirlholes.

Corrasion of the walls of rock-cut channels occurs at bends, particularly in nearly horizontally-bedded sediments, such as the Hawkesbury Sandstone of the Sydney basin, and the alternating sandstones, limestones and shales of the North York Moors, England. Whether the curved hollows close to the base flow level of the stream are developed solely by the mechanical action of shooting flow at bends armoured with rubble is doubtful. In many cases all that the mechanical action has done is to remove decomposed rock debris. However, such lateral corrasion contributes to the widening of channels and to the development of bends.

Rock particles not only act on the floor and walls of the channel, but on each other, hitting and rubbing together as they are carried downstream. Such abrasion or attrition produces a rounding of pebbles in a downstream direction and the gradual elimination of the softer rock fragments. Experimental measurements show that the *specific abrasion*, or *coefficient of abrasion*, the loss of particle

16 *Numerous swirlholes (river potholes) in the bed of the Snowy River, New South Wales. Some holes have merged. The largest boulders are about 2 m long. (Photo by J. N. Jennings).*

weight per unit weight and unit distance, of quartz is less than one-fifth that of marly limestone (Table 11). In general the rounder the particle the lower the coefficient of abrasion. Rate of abrasion also varies as the one-fourth power of the velocity, and is dependent on the character of underlying rock or particles (Graf, 1971). The relationship between channel gradient and corrasion is illustrated by the differences in the coefficients of abrasion for Austrian alpine rivers and for the lower Mississippi (Table 11).

TABLE 11 **Specific abrasion for various materials and rivers**
(after Graf, 1971)

Rock	Loss of particle weight per unit weight per km	River	Loss of particle weight per unit weight per km
Marly limestone	0.0167	Mur, Graz, Austria	0.0181
Limestone	0.0100	Gail, Carinthia, Austria	0.1054
Dolomite	0.0083	Danube, Austria	0.0230
Quartz	0.0033	Traun, Austria	0.0269
Gneiss and granite	0.0050–0.0033	Lech, Austria	0.0937
Amphibolite	0.0035–0.0020	Seine	0.0095
		Mississippi (lower)	0.00162
	-	Ohio (lower)	0.00293

Abrasion is rarely solely a mechanical process, chemical reactions are also involved. Resistant quartz fragments dominate the gravels of the Sungai Gombak, Malaysia, some 10 km downstream of the last outcrop, but the elimination of schist and granite pebbles is as much due to the selective chemical decomposition of minerals in those pebbles as to abrasion. Pebbles can also break as the result of collision with other pebbles, but as rounded pebbles with smooth edges are common in river gravel away from rock outcrops, such pebble fracture is probably a rare event.

The most spectacular corrasion forms often occur in the plunge pools at the base of waterfalls where rock fragments are swirled around by the force of the water descending over the fall. The greater energy available in such situations produces large scour and lift forces which cause corrasion of the floor and sides of the pool, sometimes undercutting the waterfall itself.

Corrasion of hard rock channels is an irregular process, affected by joints, bedding planes, chemical processes and the material supplied to the channel. Downward erosion occurs as a result, but lateral corrasion and abrasion of the sediment load account for the majority of the work expended in this interaction of water, sediment and channel bed in hard rock conditions.

Steady-state trends in the river bed-total flow-sediment system

The interactions between water, sediment and the channel bed just discussed show relationships between gradient, sediment size and velocity which suggest that if hydrologic conditions and supply of sediment from slopes remain the same for long enough, the interactions in the channel will eventually reach a state in which just as much material is being deposited in the river bed as is being eroded. Such a condition is said to be a state of dynamic equilibrium and implies that the river channel has achieved the best hydraulic condition to carry the flows of water and sediment supplied from upstream. Channels in this condition would be expected to have systematic relationships between flow, shape and sediment factors. Such relationships are expressed by the hydraulic geometry equations:

$w = aQ^b$	w = width of flow
$d = cQ^f$	d = depth of flow
$V = kQ^m$	V = flow velocity
$Qs = pQ^j$	Q = water discharge

Since, by continuity,

Other symbols are defined in

$w d\underline{V} = Q = aQ^b . cQ^f . kQ^m .$ appendix 1.

then

$b + f + m = 1$
$a.c.k = 1$

Such perfect relationships are not always found in practice. The exponents depend on channel form and the bed and bank material.

The Rio Santa Muerto (Table 12) increases in width much faster than in depth, b being considerably larger than d, an increase in depth of only 0·03 m causing a 0·76 m increase in width (Lewis, 1966). Brandywine Creek is cut into resistant material and increase in width is negligible, increases in depth being associated with increases in velocity. As the roughness effects become smaller as discharge increases, velocity increases more rapidly than discharge.

TABLE 12 **Values of the at-a-station hydraulic geometry exponents**
b, f and m for various rivers in humid lands
(after Gupta, 1975; Graf, 1971; Knighton, 1975 and Lewis, 1966)

	b	f	m	mean annual runoff (mm)
Brandywine (Penn.)	0.04	0.41	0.55	n. a.
5 Buff Bay Stations (Jamaica)*	0.11	0.55	0.37	n. a.
Bollin and Dean (England)	0.12	0.40	0.48	n. a.
Roanoke (N. Carolina)	0.12	0.47	0.41	939.8
Skagit (Wash.)	0.13	0.35	0.52	990.0
Rogue (Oreg.)	0.13	0.34	0.53	533.4
10 Rhine Stations (Germany)	0.13	0.41	0.43	n. a.
Rio Manati (Puerto Rico)	0.17	0.33	0.49	n.a.
Colorado (Colo.)	0.19	0.36	0.45	985.2
Merrimack (Mass.)	0.20	0.35	0.45	711.2
Susquehanna (N.Y.)	0.23	0.37	0.40	660.4
Big Sandy (Ky)	0.23	0.41	0.36	482.6
Sanzamon (Ill.)	0.28	0.49	0.23	457.2
White (Indiana)	0.29	0.36	0.35	635.0
Tuolumne (Cal.)	0.30	0.34	0.36	457.2
Rio Santa Muerto (Puerto Rico)	0.34	0.23	0.40	n. a.
Neches (Texas)	0.35	0.47	0.18	482.6
White (Mt Rainier)	0.38	0.33	0.27	n. a.
4 Stations Yallahs (Jamaica)*	0.42	0.32	0.24	n. a.

*Gupta (1975) attributes this contrast between rivers draining to the north and south coast of Jamaica to the rainfall regime, the Yallahs River being braided in the downstream section as a result of seasonal rainfall, whereas the Buff Bay River receives rainfall at all times of the year.

Changes in roughness are not necessarily simple power functions and thus the hydraulic geometry equations given above may only be an approximation to a more complex relationship. To cater for the channels where there are discontinuities in the change of roughness with depth, as occurs in many low order channels where base flow only occupies part of an irregular channel floor, Richards (1973) suggests that the three flow equations can be generalised into quadratic curves in logs:

$$\log w = b_1 + b_2(\log Q) + b_3(\log Q)^2$$
$$\log d = f_1 + f_2(\log Q) + f_3(\log Q)^2$$
$$\log V = m_1 + m_2(\log Q) + m_3(\log Q)^2$$

Irregular channel cross-sections may be the product of short term changes in channel dimension, such as aggradation during low

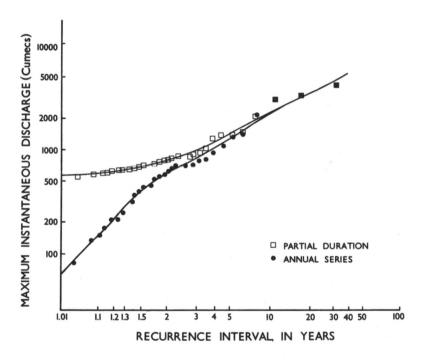

37 *The partial and annual series of flood recurrence intervals for the Murrumbidgee River at Gundagai, New South Wales, 1941–1970. Often flood analyses are concerned only with the largest flood of each year, the data for the* annual maximum *flood forming the* annual series. *However, this series ignores the second and third highest events of a year which may outrank many of the highest annual events of other years in the series. This objection is overcome by listing all events above a selected base, usually the lowest annual flood. The sample thus obtained is known as a* partial series *and is the series most appropriate for studying extreme events in fluvial geomorphology.*

flow (Harvey, 1969) but in many cases they express the way in which low flows, moderate flows, annual floods (Fig. 37) and rarer more catastrophic floods have left their imprint on the channel. Other irregularities may be due to outcrops of hard rock on one bank or the other, to the random presence of large boulders or to a recent shift in the position of the channel. When discussing hydraulic geo-metry data, it should be remembered that they are collected at regular gauging sites, deliberately chosen to have a straight uniform reach for some distance upstream. Such sites may be quite atypical of the conditions prevailing over most of the length of the channel.

Even though at any given site on a river width, depth, velocity and suspended sediment load increase as power functions or quadratic relationships of discharge, the precise relationships which deter-mine the size of the actual channel and the height of the banks are not known. Many people have suggested that the size of the channel is determined by certain large flows which occur frequently enough to keep the channel open, but which are small enough for nearly all flows to be contained well within the channel itself. Such a flow, or range of flows, is the basis of the concept of a *dominant discharge*, which implies that there is an equivalent-uniform discharge (deter-minable if the precise factors affecting it were known) which can be related to an equivalent-uniform quantity (such as width), or quan-tities, by exactly the same functional relationships as exist between their counterparts in an artificial channel with a constant steady flow (Blench, 1973). The average width of a canal designed to cope with such a constant steady flow could be estimated from the equation:

$$\frac{A}{d} = \sqrt{\frac{F_B}{F_S}} \, Q$$

where F_B and F_S are the bed factor and side factor respectively expressing the resistance of the channel perimeter. F_B can be estimated from:

$$F_B = 1 \cdot 9 \sqrt{D} (1 + 0 \cdot 12 \, C_S)$$

where D is the median bed material size by weight in millimetres and C_S the bedload charge, in parts per 100,000 of weight. F_S ranges from $0 \cdot 1$ to $0 \cdot 3$ for loams of slight to high cohesiveness (Blench, 1957). For a river, Blench (1961) suggested that the flood breadth could be calculated by substituting the peak flood discharge, Q max, for Q in the canal width equation. However, the tendency of rivers to develop meanders and sinuous channels makes such equations difficult to apply.

Nevertheless, much has been written about dominant discharge. Obviously there must be different dominant discharges for different quantities in any one river (Blench, 1973). For example the dominant discharge for breadth in a straight, incised reach of a river whose suspended load is incompetent to rebuild eroded banks is likely to be close to record flood value. On the other hand the dominant discharge affecting the slope of a reach of a channel may be the discharge which, over a period of time, would be just capable of moving the sediment carried by the river in that period. Presumably this discharge would fall between the highest flow which does not displace the bed, on the one hand, and high flood on the other; also, its water sediment complex would have to correspond roughly with that of the real river at that discharge; and the sinuosity of the river would have to be retained.

A common assertion that the dominant discharge for a wide range of natural and laboratory streams is the bankfull discharge, the discharge just contained within the banks (Ackers and Charlton, 1970), has been extended by Dury (1969b) into the notion that:

dominant discharge = bankfull discharge = discharge at the recurrence interval of 1·58 on the annual flood series (i.e discharge at the most probable annual flood).

This notion is supported by analyses of processes such as sediment delivery which show that channels are shaped not by events of great magnitude and low frequency but by those of modest magnitude and high frequency, and by the water at bankfull stage being in contact with and acting upon the whole perimeter of the channel. However, other evidence suggests that discharges less than bankfull can control channel shape (Carlston, 1966), while in England, boulder clay channels seem to be related to mean annual flood discharges (recurrence interval 2·33) and chalk catchment channels to less frequent events (Harvey, 1969). Many Australian rivers appear to have two or more sets of channel banks, or benches within channel. At some sites the present active floodplain may be a relatively inconspicuous, narrow, depositional bench below the apparent floodplain, possibly because the stream channel has been incised as a result of changed hydrologic conditions. Below the floodplain level benches related to seasonal high discharges or to more exceptional events may occur. Woodyer (1968) found that bankfull discharge at the middle bench level in a group of New South Wales rivers occurs once in every 1·02 to 1·21 years, approximately the normal wet season flood level. Bankfull discharges at the high bench or active floodplain level recur once every 1·24 years or more (Table 13).

Despite the quantitative relationships now available for a wide variety of gauging stations in different climates, the true nature of the channel forming or dominant discharge is not understood. Qualitatively, the influence of discharge, sediment load and channel perimeter characteristics is readily assessed by examining the form of the channel and assessing how many of the channel types recognised by Tricart (1960) apply. Tricart suggested four types which could relate to three benches and a floodplain.

The *base flow channel* occupies the floor of the channel between vegetated banks, and carries the lowest flows, which often wind about through the gravels or sands of the minor channel. Frequently the base flow channel takes the form of a series of pools linked by shallow riffles over and between coarse gravels. At the lowest flows, the pools may be almost stagnant, linked only by seepage through the channel bed material. The line through the deepest points of the base flow channel, the lowest points of the wetted perimeter, is termed the *talweg* (Fig 38).

The *minor channel* is generally clearly defined by steep almost continuous banks which rise from the channel floor. Flows occupying the whole of the minor channel occur sufficiently often to prevent the growth of vegetation. The upper parts of the banks, however, have a grass cover with shrubs or rapidly growing trees. The lower bank, usually part of the wetted perimeter, may be distinguished from the upper bank, normally above water and vegetated.

The *periodic major channel* is occupied by floods occurring at least once a year. This gives rise to specific ecological conditions, where only those plants that can withstand submersion grow. In western Europe, willows and alders survive within the periodic

TABLE 13 **'Bankfull' stage of some New South Wales rivers**
(after Woodyer, 1968)

River	Gauging site	Floodplain stage (metres)	Return period (annual flood series)
Dumaresq	Roseneath	8.1	7.9
Macintyre	Mungindi	7.3	3.9
Macintyre	Mogil Mogil	6.7	1.63
Barwon	Walgett	8.5	1.36
Darling	Bourke Town	11.5	3.54
Darling	Louth	10.2	4.6
Darling	Wilcannia	7.1	1.54
Peel	Piallamore	4.4	2.02
Peel	Carrol Gap	4.5	1.59
Moonan Brook	Moonan Brook	3.1	8.5
Murrumbidgee	Yaouk	2.1	1.61
Mangarlowe	Marlowe	7.1	6.59

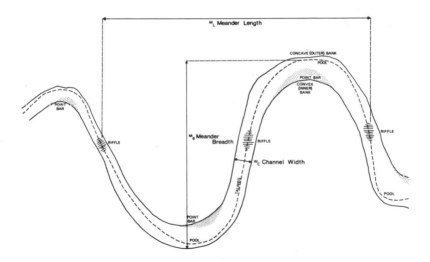

38 *Channel forms in a meandering stream showing the definitions of the major parameters used in hydraulic geometry studies. Meander length (M$_L$) is measured from the point of inflection, or crossover, across the complete meander wave to the next point of inflection. Meander breath (M$_B$) is measured from the deepest point in the talweg at the outer limits of the meander curves. Channel width (W$_c$) is the distance between the defined, usually vegetated, banks of the channel, but in some analyses, the width of the area wetted by the flow of a given recurrence interval is appropriate.*

channel, while in humid tropical regions, the periodic channel may be characterised by a shrub or grassy vegetation, or by the absence of certain tree species found in dryland forests. In traditional agricultural economies, the periodic channel is used for pasture rather than cultivation, for example for grazing by buffalo in the padi areas of south-east Asia.

The *exceptional major channel* is occasionally used by major floods. So infrequent are flows filling this channel that the vegetation differs little from that of the interfluve areas outside the flood channel. The extent of this exceptional major channel is often not recognised; the frequent flooding of such New South Wales towns as Singleton, Maitland, Kempsey and Grafton shows how settlements have inadvertently been sited within the bounds of the extreme flood channel.

Tricart's classification qualitatively illustrates how the mutual interaction of bed, flow and sediment transport create a variety of forms and how an individual cross-section may express the geomorphic role of several orders of hydrologic event. However, the variation of channels with slope and their tendency to sinuosity produces the variety of channel shapes and patterns discussed in the next chapter.

VII

THE SHAPES AND PATTERNS
OF RIVER CHANNELS

Just as river channels adjust their cross-sections to cope with varying flows of water and sediment, so they adjust their slopes and their channel patterns. The adjustments between slope, width, depth and discharge give the channel three degrees of freedom, while a fourth is available from the tendency to sinuosity, to the development of meanders. The slope of the channel from source to mouth is the longitudinal profile, while the plan of view of the channel is the channel pattern.

THE LONGITUDINAL PROFILE

As soon as water begins to flow along a natural channel it starts altering the shape of that channel both in cross-section and in gradient. If water is poured on to a heap of loose sand, it rapidly washes sand off the steep slopes of the heap and spreads it out in fan-like deposits at the base of the heap. The channels occupied by water running over the sand are continuously adjusted to more gentle gradients, as long as the flow of water on to the sand remains constant. The channels may thus be said to evolve towards an equilibrium between the force of the water flowing on to the sand and the resistance offered by the sand.

This simple example indicates how natural channels tend to erode the steepest parts of the courses and build up the gentle parts to a single, uniform gradient. The notion that a stream tends to equalise its work in all parts of its course was developed by Gilbert (1880) from studies of land sculpture in the Henry Mountains of Utah. As the corrasive and transporting power of a stream depends on the downslope component of gravity, such power expresses itself in the contrast in channel longitudinal profile form through materials of varying resistance. In hard rocks, many metres of channel fall are concentrated into short horizontal distances, while in soft materials channels fall steadily over long distances. The quantity of material being carried by a stream affects its power to corrade the channel bed. If the stream is carrying all the material it is competent to

135

transport, there is no corrasion of the channel. If it has less than a full load, it takes up more by the corrasion of its bed.

The long profile of a river tends to acquire a smooth shape through the filling of declivities and wearing away of abrupt sections. The present shape of the longitudinal profile of any river or stream depends on the geological history of the drainage basin, as well as on the climate of the area. Many large rivers appear to have attained a profile approximating to a smooth curve, concave upwards towards the sky (Birot, 1966), as is the case with the Sungai Kelang in West Malaysia (Fig. 39).

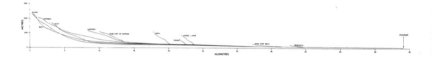

39 *Long profile and valley-side profiles of the Sungai Kelang, Malaysia*

Rivers ranging in size from small badland gullies to the Nile all have profiles showing some degree of concavity, but some have convex portions (Leopold, Wolman and Miller, 1964). Even though rivers such as the Nile and Rio Grande flow through arid regions in their lower reaches, and lack an increase in discharge, they have concave profiles indicating that the normal decreasing gradient downstream depends on more than increasing discharge. As rills developed over one or two years on newly cut slopes have similar profiles, age by itself cannot determine concavity, which must be the product of several interacting factors.

The detail of longitudinal profiles

The many irregularities and changes of slope in longitudinal profiles are illustrated by the River Adour in south-western France, which above the confluence with the Arros has a mean slope of 1 in 500 (range 1 in 250 to 1 in 1000), just at the lower limit for a torrential stream (Vogt, 1962). From the Arros confluence to Saint Sever the mean slope is 1 in 770 (range 1 in 400 to 1 in 10,000), while below Sever it is approximately 1 in 2000 (range 1 in 1000 to 1 in 5900). Even in the lowest section, the gradients are steep for a river flowing over the surface of a plain and probably account for the unstable river channel which suffers lateral erosion of 5 to 17 m yr^{-1} and frequent changes of course.

More marked variations of gradient occur where streams flow over rocks of differing resistance to erosion, as on the River Nile where the Shabulka Gorge and five cataracts below Khartoum mark the points where the river leaves the sandstones and crosses resistant crystalline rocks. Major waterfalls, such as the Victoria Falls on the Zambesi, or High Force where the River Tees drops over the vertical face of the Whin Sill of northern England, are caused by belts of igneous rock. The humid lands of the Eastern Highlands of Australia are marked by several falls over the edge of basalt flows, such as the Millstream Falls on the edge of the Atherton Tableland and the Ebor Falls on the Dorrigo Plateau.

Not all variations in the declivity of the longitudinal profile may be accounted for by contrasts in resistance of rocks over which a river flows. The longitudinal profile of the North Tyne River steepens markedly below Bellingham, England, causing the smooth concave upwards profile of the section upstream of Bellingham giving way to a steeper, irregular profile (Peel, 1941) as a result of rejuvenation of the drainage system by a fall in sea level.

Waterfalls and rapids

While waterfalls may be regarded as interruptions of the long

17 *Millaa Millaa Falls over the columnar basalts of the Atherton Tableland, north Queensland.*

profiles of rivers, they are often the most impressive fluvial landforms and the aspects of rivers of most interest to the general public
for their value as either scenic attractions or as sites for hydroelectric power generation. A waterfall is a sharp break in the long
profile of a river, where the water drops for a vertical distance in
one fall. A cataract is a step-like succession of waterfalls, while
rapids are sections of the long profile in which the flow of water is
broken by short vertical drops which seldom affect the complete
width of the channel at any one cross-section. A cataract or waterfall is never drowned by an exceptional flood, but rapids may be
entirely inundated by high flows, whose turbulent water surface may
be the only indication of the underlying rapids.

Schwarzbach (1967) has put forward a genetic classification of
waterfalls which distinguishes those that have developed indepen-
dently of the erosive activity of the river from those that have
developed as a result of such activity. The two classes of waterfall
may be termed river independent and river dependent. A further
distinction must be made between destructive waterfalls in which
the waterfall is being worn away and constructive waterfalls, where
the rock of the fall is being deposited and the fall is advancing downstream.

From Schwarzbach's classification (Table 14) emerge three over-
riding genetic factors:

a) the importance of a wet climate;
b) the significance of tectonic uplift and the legacy of glaciation;
c) extensive layers of resistant cap rocks (such as the Hawkesbury Sandstone) which form the ledges of waterfalls.

The highest and biggest falls generally occur on the uplifted
edges of old land surfaces, as in the great falls of Surinam, Guyana
and the Eastern Highlands of Australia.

Rates of waterfall evolution

Between 1890 and 1905, Niagara Falls retreated at a rate of 1·64
m yr^{-1} (de Martonne, 1951), while the Shablinka River waterfall
near Leningrad has been retreating 0·33 m yr^{-1} (Selivancy and
Svarichevskaya, 1967). Both these falls have a similar structure
with nearly horizontal limestones being the resistant fall-making
strata with weaker shales below being undercut by abrasion on the
rear wall of the deep plunge pool below the falls. At Niagara the

TABLE 14 **Schwarzbach's genetic classification of waterfalls**

A. *DESTRUCTIVE FALLS* (with more or less distinct headward erosion)
   I *CONSEQUENT FALLS* The river follows a 'natural surface' and the falls
      are located at pre-existing breaks in the river profile. Such breaks are by
      and large independent of the river's own erosion. The subsurface structure
      may be completely homogeneous.
      1 Falls caused by the sudden blockage of a valley or by any other deviation
         imposed on the talweg
         a Falls at the downstream edge of the blockage (landslide, lavastream,
            etc.) e.g. Godafoss
         b Rapids over large boulders
         c Falls at a point where a river rejoins its old course beyond the deviation,
            often on the old valley side, e.g. Saale near Bernburg, 1933
         d Falls located at a new junction point with the main valley following
            a blockage, e.g. Thvera
         e Falls following the cutting off of a meander, e.g. Lauffen on the Neckar

   II *SUBSEQUENT FALLS* No primary break of the river's talweg is visible
      but the potential for the break is there in the inhomogeneity of the subsurface
      geology. The river seeks it out. The subsurface flow may exert its influence
      from the beginning of the stream flow ('primary subsequent falls', which are
      stationary), or a consequent fall may work headward until its original
      significance is lost ('secondary subsequent falls').
      1 Horizontal or gently dipping (hard and soft strata). The majority are
         secondarily subsequent with headward erosion, e.g. Great Niagara,
         Gullfoss; waterfall steps.
      2 Excavation of a steeply dipping boundary between hard rocks upstream
         and soft rocks downstream. 'Primary' subsequent falls with little if any
         headward erosion. The fall is often stationary.
         a The sharp boundary is volcanic in origin (vertical barrier falls), e.g.
            Yellowstone River Falls, Baulufoss
         b The boundary under attack is a fault, e.g. Montmorency Fall near
            Quebec
         c The boundary is erosional in origin (vertical barrier falls), e.g. Rhine-
            fall near Schaffhausen
      3 A joint zone is eroded and headward erosion is bound to the joints,
         e.g. Victoria Falls, Iguacu Falls

B. *CONSTRUCTIVE FALLS* (Downstream movement and no headward erosion.)
   The break is usually the result of $CaCO_3$ precipitation.
   1 $CaCO_3$ precipitation at the mouth of a secondary valley or on a slope,
      e.g. Uracher waterfall, Cascata della Marmore, Pliwa Falls
   2 Overflow of waters penned in a valley lake dammed directly as the result
      of travertine deposition, e.g. Lakes of Plitvitz

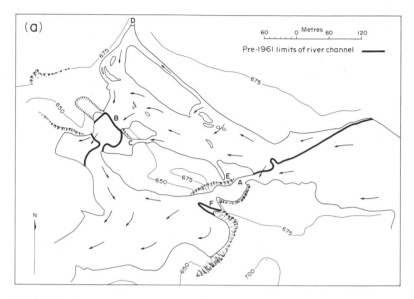

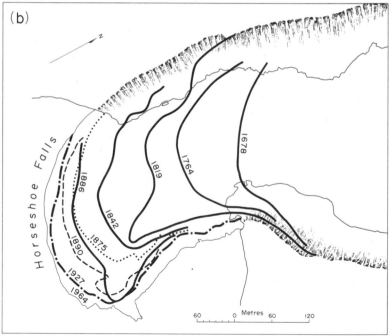

40 *Evolution of (a) the Murchison Falls, (b) the Niagara Falls (after (a) Ware, 1967 (b) Philbrick, 1970).*

water and entrained rock debris strike the base of the 50 m high fall at about 220 m s^{-1}. Waterfall recession rates are by no means constant. Comparisons of surveys from 1842 to 1966 show that recession is faster when the Horseshoe Falls have a well-defined horizontal notch in the crest, being slower when the crest is in the form of a horizontal arch (Fig. 40). The horizontal arch crest is thought to be the more stable waterfall configuration (Philbrick, 1970). The Gullfoss, in basalt and conglomerate in Iceland, retreats 0·25 m yr^{-1} (Schwarzbach, 1967).

The Gullfoss, Niagara and Shablinka Falls are all in areas glaciated during the Pleistocene and their rivers carry adequate coarse debris to abrade the falls. By contrast, falls in tropical rivers seem to be much more permanent, possibly because many tropical rivers carry little bed material coarser than sand size, Tricart (1959) finding only one pebble over 5 cm long in over 200 swirlholes in the Felou Falls on the Senegal River. The absence of coarse debris in such rivers reduces their abrasive action. Resistant fall-making strata are subject to less attack than the rocks in temperate streams loaded with relict glacial debris. In some tropical waterfalls and rapids, however, many pebbles can be found, the rapids of the Sungai Dong in Pahang, Malaysia, having many swirlholes containing pebbles 4 to 6 cm in diameter. The Dong, however, is much smaller than the Senegal and is supplied with pebbles from quartzitic sediments upstream. Usually, pebbles are only transported short distances in tropical rivers before being reduced in size by chemical weathering and abrasion (Michel, 1973).

The role of structure in the evolution of tropical waterfalls is well illustrated by the evolution of the Murchison Falls on the Victoria Nile in Uganda. Before 1961, the whole of the discharge of the Victoria Nile passed through a joint-controlled 6 m gap (marked A on Fig. 40a). The line of the joint slot continued north-eastwards across the former line of the river and caused the river to take a sharp turn towards the west. The water gathered momentum on entering the slot to descend the modest 42 m of the fall.

At the end of 1961 the discharge of the Nile increased suddenly owing to a general increase in precipitation and lake levels in East Africa. Higher water levels in the Nile eventually led to the re-adoption of the former course of the river, over the joint step and round to a formerly dry cascade (Bishop, 1967). The river now flows round a bend some 0·8 km long over a new fall (marked B on Fig.

40a) with a lip some 10 m lower than the old joint slot. Murchison
Falls now comprises three separate cascades (A, B and C on Fig.
40a), which are probably similar to those described in a 1907 survey
(Ware, 1967). The joint slot that dominated the falls in the decades
before 1961 had probably not come into existence when Sir Samuel
Baker explored the Lake Albert basin in 1864. These changes in the
Murchison Falls over a few decades illustrate the interaction of
structure and hydrology in the development of streams.

Another factor, dependent on structure and hydrology affecting
waterfall evolution, is geochemistry. Although rocks of valley-side
slopes are usually subject to active chemical weathering, those exposed in the channel are often highly polished, sometimes carrying
an extremely resistant 2 to 20 mm thick iron-rich silicate patina of
dark brown, violet or blackish colour (Tricart, 1959). The extent of
such patina varies with the position of the rock surface in relation
to the low flow, minor and major channels of the river. Where the
rock is regularly wetted by storm flows, yet dry at base flow, such
superficial encrustations are most likely to develop. Iron and man-
ganese layers cover the rocks of soelas, or rapids, of the Corantijn
River in Surinam (Zonneveld, 1969). The slow evolution of many
tropical waterfalls suggests origins well back in the Tertiary
(Schwarzbach, 1967), Büdel (1957a) finding evidence of great age
in distinct varietal differences between water snails at each cataract
on rivers in Surinam.

By contrast, in unconsolidated materials, the cutting back of a
waterfall or nick point may be rapid, a flume experiment producing
the elimination of a nick point by 8·5 m of recession in 26·7 hours
(Brush and Wolman, 1960). During short periods of intensive erosion, small gullies can cut back several metres per day (Leopold,
Wolman and Miller, 1964). Waterfall evolution and nick point
recession are thus part of the total complex of river channel development dependent on the discharge regime and the materials into
which the river channel composition is cut (Plate 18). Changes in
discharge, such as those at the Murchison Falls, are as crucial to
waterfall evolution as changes in the stability of the fall-making
rock, such as those at Niagara.

*Channel instability: rejuvenation, climatic change and profile ir-
regularity.*

In many streams, such as the River Thames in England, small

18   *Headward retreat of a miniature waterfall. A hardpan acts as a cap rock,
     but has broken off repeatedly after the underlying clay loam has been
     washed out. Near Palmer, South Australia. (Photo by J. N. Jennings)*

islands, aits, occur where the channel divides into separate arms for
a short distance. Such subdivision of the channel may be associated
with breaks in the longitudinal profile due to rejuvenation (Waters,
1953) or with climatic change, as in the anastomosing channels of
the Amazon and Orinoco.

On the Rio Caroni, a tributary of the Orinoco, structural
influences and climatic change have produced a complex series of
waterfalls, rapids, active and abandoned channels. The headwaters
of the Caroni rise on the resistant horizontal Precambrian rocks of
the Roraima Series of the Venezuelan-Brazilian shield. Near the
edge of the Roraima Plateau the streams become incised in
V-shaped, notch-like clefts, before plunging over waterfalls, including the 1000 m high Angel Falls. The plateau surface probably
remained humid throughout the Pleistocene and is still covered with
rain forest. Below the falls the Caroni has an anastomosing channel
system, developed from an original series of ephemeral sheet floods
under arid conditions into a pattern of incised channels which were
subjected to stream piracy as certain channels developed in pre-
ference to the others. Much of the present-day flood discharge of

the Caroni flows under vegetation, with sometimes a virtually continuous sheet of water flowing through a wide area of vegetation (Garner, 1966). These types of channel pattern are the result of alternating arid-humid conditions and are not usually found in areas that have been continuously humid.

Causes of longitudinal profile concavity

The longitudinal profile of a stream is a graph of distance versus elevation and may be considered to be a function of the following variables, changes of each of them in the downstream direction affecting the shape of the profile:

Q	discharge
Q_T	total sediment load delivered to the channel
D	particle size diameter (calibre of debris)
V	velocity
w	width
d	depth
s	channel slope

If three theoretical profiles, one straight, one moderately concave and one markedly concave are considered, it can be seen that the straight profile would require an increase of velocity downstream equivalent to the downstream increase in depth, conditions which seldom occur in natural channels. Depth usually increases downstream more markedly than velocity. In a moderately concave stream, the case typical of most natural rivers, depth and width increase markedly downstream, but velocity decreases slightly. However, should width be held constant, depth would increase greatly with a slight increase in velocity.

The rare cases of streams with convex profiles are probably the result of something other than approximately equal rates of downstream increase in depth and velocity. In semi-arid environments, downstream decrease in discharge may occur, due to seepage into the ground or evaporation losses from the stream.

The relationship between downstream decrease in bed material size through abrasion and channel gradient described in the previous chapter affects profile shape; the more rapid the downstream decrease, the more concave is the longitudinal profile. When streams of the same drainage area are compared, channel slope is related to median bed-material size, decreasing downstream as the 0·6 power of median size (Hack, 1957).

Regime and graded rivers

While in terms of geological time (Tricart's order 1 to 5, Table 1) river channels must be regarded as part of the fluvial denudation system undergoing continued change, in shorter time spans the hydraulic geometry process-response system is composed basically of independent variables, water and sediment discharge, and the variable channel morphology. This process-response system is either in or adjusting towards a temporary equilibrium. This adjustment through time must eventually lead, if conditions remain unchanged, to the final steady state of the longitudinal profile, a smooth concave upward curve. As explained above, many factors intervene to prevent the attainment of this ultimate form, even if such a curve may be expected to be the ultimate profile form for all rivers. Nevertheless, streams may display virtually constant channel form and gradient over extended reaches. Such streams have been described as in regime or graded (Santos-Cayade and Simons, 1973). Peel (1941), for example, defined the profile of the North Tyne upstream of Bellingham as a graded profile, in that all the irregularities due to lithology, tectonics or change in altitude of the river mouth have been eliminated.

Regime applies to the cross-section, longitudinal profile and channel pattern, Blench defining it as meaning that the 'average values of the quantities considered as constituting regime (width, depth, velocity, meander pattern, water and sediment discharge, bed material and slope) do not show a definite trend over some interval of time'. Mackin (1948) defines a graded stream as one 'in which, over a period of years, slope and channel characteristics are delicately adjusted to provide, with available discharge, just the velocity required for the transportation of the [sediment] load supplied from the drainage basin'.

To obtain the conditions required by these definitions, the independent variables, water and sediment discharge, must constitute a stationary time series, that is, if the minor variations between low flows and floods, and from one flood to another, are smoothed out, there must be no trend towards increasing or decreasing water or sediment load. Such stationary time series may exist for a few decades, but there are few records of longer term stability of these independent variables. The concept of regime is thus useful in looking at short term changes to a reach of a river, especially the type of change that would be considered in an environmental impact study.

The notion of the graded stream is difficult to apply in reality, Garner (1974) stating that a protracted search for streams in the graded state has been fruitless, though short reaches of various rivers have been said to be in quasi-equilibrium. Dury (1966a) found that 'because the concept of grade necessarily involves the converse idea of an ungraded state, it is to be recognised as unserviceable both in the study of actual terrains and in the theoretical analysis of land-forms generally'.

Nevertheless, the concept of regime and a trend towards an equilibrium condition are extremely valuable in trying to explain why rivers develop their shapes and patterns. From such considerations Yang (1971a) has developed two basic laws of stream morphology. The *law of average stream fall* states that for any river basin that has reached dynamic equilibrium condition, the ratio of the average fall between any two different order streams in the same river basin is unity. The *law of least time rate of energy expenditure* states that during the evolution towards its equilibrium condition, a natural stream chooses its course of flow in such a manner that the time rate of potential energy expenditure per unit mass or weight of water along this course is a minimum. One consequence of this law is said to be a requirement that the channel slope decreases in the downstream direction in order that the *unit stream power*, the time rate of energy expenditure per unit weight of water, has a value of zero when the river reaches the sea. This requirement explains why the longitudinal profile is usually concave (Stall and Yang, 1972).

POOLS AND RIFFLES

Although rivers characteristically develop a uniform profile down-stream, gradually decreasing in steepness, they also tend to maintain an alternation along the talweg of low-gradient deep pools and higher gradient riffles or rapid reaches. At the side of the channel by pools, coarse material accumulates on point bars. A *pool* (Fig. 38) is a topographically low area produced by scour which generally contains relatively fine bed material. A *point bar* is an accumulation of relatively coarse bed material on the concave side of the talweg adjacent to a pool. The point bar and the pool together produce an asymmetrical cross-profile. A *riffle* is a topographic high area produced by the lobate accumulation of relatively coarse bed material (Keller, 1971). The inflection point of the talweg as it swings from one side of the channel to the other (Engels. 1905) is located on the riffle approximately halfway between successive pools. The cross-profile of the riffle is generally symmetrical.

A range of particle sizes is necessary for pool and riffle development, channels with uniform sand or silt having little tendency to form pools and riffles. This may be related to the shallowness of sandy channels and the lack of high cohesion, with a consequent trend to braiding (Leopold, Wolman and Miller, 1964).

In streams that are cut into bedrock. similar sequences of pools are found, although the pools may have irregular shapes determined by joints in the bedrock. In the Hawkesbury River downstream of Windsor, New South Wales (Dury, 1970a), and its tributary, the Colo (Dury, 1966b), irregular rock floored pools occur where joints of the thickly-bedded Hawkesbury Sandstone have provided primary paths for water erosion of bedrock. Once water has enlarged the joints, the sandstone may disintegrate into particles of varied sizes which will be washed downstream 'at times of high discharge. Irregular bed profiles of pools may also rise from the slipping of large blocks into the river, as may have happened in the pool near the hill Wheelbarrow on the Colo River (Dury, 1966b).

The spacing of pools and riffles in New South Wales streams at 4·8 to 7·1 times channel bed width (Table 15) corresponds to North American (Langbein and Leopold, 1966) and British (Harvey, 1975) findings of spacing at 5 to 7 times bed width. Even in the Grand Canyon section of the Colorado River, rapids, equivalent to riffles in smaller streams, are regularly spaced at intervals of 2·6 km

TABLE 15 **The spacing of pools and riffles in New South Wales streams**

River	Pool spacing (metres)	Bed width (metres)	Spacing: width ratio	Source
Colo	535	75–110	4.8–7.1	Dury, 1966b
Hawkesbury	1020–1580	721–310	5.9–5.1	Dury, 1967
Port Hacking	60	10	6.0	Shoobert, 1968
Oxford Creek	27–30	4.7	5.8–6.3	Kelly, 1972

(Leopold, 1969). This occurrence of rapids, separated by deeper pools, is apparently independent of the major bedrock type and the valley characteristics associated with bedrock types. The pool and rapid succession appears to be one aspect of channel adjustment towards maintaining stability, or quasi-equilibrium, and is typical of rivers of all sizes.

The rapids of the Grand Canyon account for most of the fall of the river through the canyon; 50 per cent of the total decrease in elevation takes place in only 9 per cent of the total river distance. Thus although the longitudinal profile of the river shows a steady fall in level, in detail the bed is irregular. Detailed profiling of the bed reveals that the pool and rapid succession in the Grand Canyon is an almost permanent feature, unlikely to have changed much over long periods of geological time. Two surveys of the Hawkesbury River downstream of Windsor (Dury, 1970a) show that between 1870 and 1969 there has been no change in channel form, although considerable man-made changes have occurred upstream of the surveyed reaches. The permanence of riffle or rapid position and spacing arises because maintenance of channel stability requires a river bed consisting of alternating deeps and shallows which remain in a constant position (Leopold, 1969).

Origin of rapids and pools in hard rock channels

Rapids in the Grand Canyon arise from the fall of large blocks into the river from adjacent high cliffs; the outcrop of hard rock layers; and the tendency for gravel accumulations to develop in all rivers. The first two causes would not account for the regular spacing of pools and rapids; only the sorting of bed material into riffles or gravel bars would do so. The development of pools below rapids is similar to the enlargement of plunge pools below waterfalls. Immediately below a steep rapid, a large part of the downstream flow will be thrown against one bank, particularly if that bank is a vertical cliff. When this occurs, the opposite side of the stream will invariably have a strong upstream current at the water surfaces, often forming half the total stream width. The shear zone between the downstream and upstream currents is characterised by surface water boils, the expression of upward water movement. To have such vertical motion there must also be downward currents. Such downward vertical motion occurs at the foot of rapids, where the deep holes and pools represent scour by downward directed water, much of which must flow along the bed at high velocities, later to appear further downstream in the surface boils (Leopold, 1969). Such effects are likely to occur in most steep rock channels, but little information is available on flow velocities and bed forms in such streams. Echo-sounding of channel bed topography may not always distinguish rock-cut, large boulder and gravel accumulation bed forms. A proportion of rapids in hard rock streams is likely to arise through transport and sorting processes similar to those which operate in alluvial channels.

Origin of riffles and pools in alluvial channels

Langbein and Leopold (1966) used experimental evidence to show that interaction among the individual units moving in a continuous flow pattern causes velocity of the units to vary with their distance apart. Concentrations of these units have certain properties encompassed under the term 'kinematic waves'. With these properties it is unlikely that such particles can remain uniformly distributed for any distance along the flow path, for random perturbations lead to the formation of groups or waves which take the form of pool and riffle sequences in gravel river beds. The spacing of the riffles is related in part to a thin veneer of gravel set in motion by the flow. The accumulation of coarse material represents an interaction between two opposing factors: increasing water velocity, which tends to increase wavelength, and decreasing amplitude, due to erosion, which tends to decrease wavelength. This balance results in riffle bars which do not move appreciably downstream.

The development of random perturbations simply through the kinematic wave properties of particles in motion does not satisfy all geomorphologists, Tanner (1968) suggesting that the sinuosity of the talweg and the pool and riffle profile may be related to the presence of roughness elements such as gravel patches in the channel walls. Just one rough patch is considered sufficient to produce oscillatory motion for a long distance downstream.

The formation and maintenance of pools and riffles appear to be related to a range of discharges up to the periodic major channel bankfull stage, above which drowning out of the effect of bed topography on water surface slope tends to occur. Although riffles may be formed primarily by flows exceeded only 10 to 20 per cent of the time, when coarse gravel is in motion, the movement of smaller material at more frequent discharges is important for the maintenance of the pool and riffle sequence (Harvey, 1975).

At an asymmetrical channel cross-section through a point bar and a pool, local velocity gradient, mean velocity, mean boundary shear stress and stream power all decrease from the deeper outer side of the channel towards the inner bank. Particles tend to be pulled from the outer bank and to cross the bed to join many more particles derived from further upstream to form a point bar on the inner side. Other particles from the outer are carried towards the talweg and downstream to be added to the upstream edge of a riffle. Such velocity distributions and sediment movements are analogous to the processes which occur in meandering streams, suggesting that the two phenomena may have a joint cause. According to Yang (1971b) river meandering is the lateral adjustment, and pool and riffle formation the vertical adjustment, of a natural river to minimise its time rate of potential energy expenditure per unit mass of water along its course of flow. Leopold and Wolman (1957) show that wavelengths of meanders and riffles form a common series which may be correlated with periodic major channel bankfull discharge and channel width.

MEANDERING CHANNELS

Regime theory, described earlier, sees the tendency towards sinuosity in rivers as part of the general tendency for channels to adjust to a form where there is a uniform distribution of forces acting on the channel bed and banks and of the retarding effects exerted by the bed and banks on the flowing water body (Langbein and Leopold, 1966). Sinuosity, the meandering tendency, is usually measured by the ratio of stream length to valley length. A straight stream has a sinuosity of 1·0, the value increasing as a stream departs from a straight line. Five classes of river channel sinuosity; straight, transitional, regular, irregular and tortuous (Fig. 41) are suggested by Schumm (1963b), who found that with decreasing sinuosity (H), the width-depth ratio of the channel (F) increases, the weighted mean per cent silt clay in the perimeter of the channel (H) decreases and the mean annual discharge increases. Thus from Schumm's empirical data:

$$P = 0.94 \, H^{0.25}$$
$$F = 255 \, H^{-1.08}$$

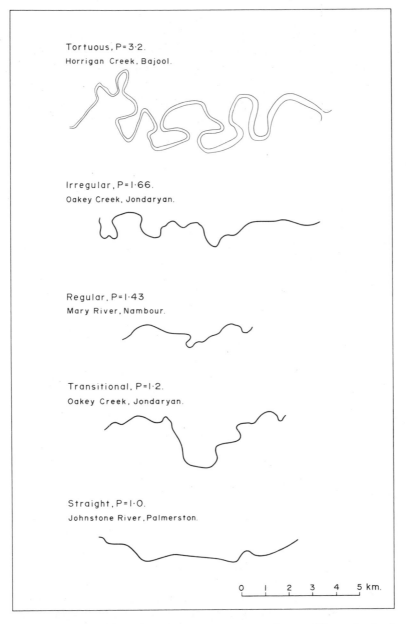

Tortuous, P=3·2.
Horrigan Creek, Bajool.

Irregular, P=1·66.
Oakey Creek, Jondaryan.

Regular, P=1·43
Mary River, Nambour.

Transitional, P=1·2.
Oakey Creek, Jondaryan.

Straight, P=1·0.
Johnstone River, Palmerston.

0 1 2 3 4 5 km.

41 *Sinuosity of various Queensland streams according to Schumm's classification*

Meandering channels occur on glacier surfaces, the Gulf Stream (Leopold and Wolman, 1960), alluvial material, solid rock and limestone lapies (Zeller, 1967). Meandering may occur in any Froude and Reynolds number range, and may be accompanied by the existence or absence of bed load transport. For all types of meanders the following empirical relationships of meander length (M_L) and meander width (M_W) to channel width (w_c) have been found:

Leopold and Wolman (1960) $M_L = 10 \cdot 9 w_c^{1.01}$

                            $M_W = 2 \cdot 7 w_c^{1.01}$

Zeller (1967) $M_L = 10 \cdot 0 w_c^{1.025}$

                            $M_W = 4 \cdot 5 w_c^{1.00}$

As channel geometry is related to some dominant discharge, an equation to predict meander wavelength from a possible channel-forming discharge may be developed. Data from three stations on the Murrumbidgee River, Australia, and thirty-three stations on the Great Plains of the western United States (Schumm, 1967) yielded the relationship:

$$M = \frac{234\,Q_{ma}^{0.48}}{H^{0.74}}$$

where Q_{ma} is the mean annual flood (recurrence interval 2·33 years).

The origin of meanders

While meandering has been ascribed to such factors as the rotation of the earth, maximisation or minimisation of energy loss, local disturbances, bank erosion, the sediment load of the stream, the development of helicoidal flow, and maladjustment between the hydraulic properties of the flow and channel shape, the basic physical character of stream behaviour has not been fully explained. Flume experiments on an alluvial channel showed that immediately after the introduction of water and sediments at the upstream end of a prepared trapezoidal channel, the bed developed ripples, symmetrically placed about a longitudinal line (Ackers and Charlton, 1970). After a period of 6 to 24 hours, depending on discharge, the symmetry deteriorated and shoals began to form near the sides of of the channel along its length. These were at fairly regular intervals, the shoals forming on alternate sides of the channel with deeps near the opposite bank. The shoals continued to increase in length and height while they migrated downstream. Then, almost simultaneously along the channel, embayments were eroded in the banks opposite each shoal. These expanded, producing a sinuous channel pattern in plan, and also migrated downstream, although more slowly than the shoals had done. The pitch of these features then remained constant.

This change from a straight to a sinuous channel arises because in the straight reach, even at bankfull stage, the alternation of steep water surface slope over riffle and flat gradient over the pool is not eliminated. This variation of shear and friction is minimised as the channel develops a meander pattern. In channels in which pools and riffles occur, meandering is a more stable geometry than a straight or non-meandering channel pattern (Langbein and Leopold, 1966).

These empirical observations of meandering developing as the stable form of a single thread alluvial channel support Yang's claim (1971c) that meandering is the only possible stable unbraided natural channel form. A single thread channel may follow a straight, zigzag or meandering line course, but to minimise the unit stream power, a straight channel between two points would need to increase its length. A zigzag or meandering path between the two points would provide the same unit stream power, but the straight segments of the zigzag channel would be as unlikely to exist as the overall straight channel. The meandering channel is the only possible stable pattern.

Meander evolution

Meanders migrate laterally and downvalley. The 5·6 m wide meander channel of Watts Branch near Rockville, Maryland, cut 3·2 m laterally in six years (Leopold and Wolman, 1960), while runoff from the catastrophic hurricane Camille in Virginia in 1969 eroded 7 m of bank in 7 m wide streams in a single flood (Williams and Guy, 1973). Migration rates vary with all the factors of hydraulic geometry, but meanders tend to increase their curvature as they evolve until cutoff occurs and the channel length is shortened and steepened. Sundborg (1956) has argued that a meander loop must of necessity widen until the radius of curvature has become so large that a further increase would slow down the flow velocity to such an extent that lateral erosion would cease.

Meandering rivers have sections affected by recent cutoffs, areas of rapidly eroding meander loops and relatively stable reaches

(Plate 19). Some show elements of both meandering and channel subdivision by gravel bars (Knighton, 1975). Probably the best examples of free meander development are the lowland streams of the humid tropics, such as those flowing to the southern coasts of Kalimantan, Irian Jaya and Papua. One of these streams, the Angabunga of Central Papua, has a sinuosity of 1·8 to 2·3 but a variety of meander wavelengths which probably arise from the irregular and rapid manner in which channel changes, cutoffs and point bar development occur (Speight 1965a, b). Application of spectral analysis to meander wavelengths has illustrated this com-

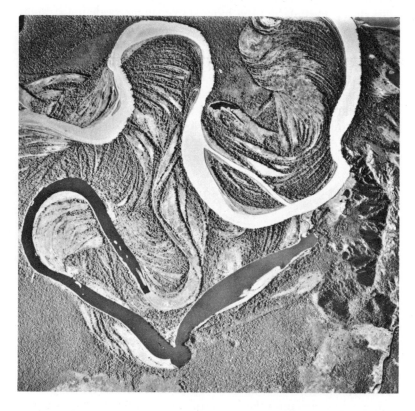

19 *Sepik River near Ambunti, Papua New Guinea, showing active meanders, chute cutoff, scrolls, point bars and an oxbow lake. (Crown Copyright reserved. By courtesy of the Director of National Mapping, Commonwealth of Australia.)*

plexity but as yet not demonstrated the most probable meander patterns for particular environments (Speight, 1967b; Scheidegger, 1970). No orderly downstream migration of meanders was found in the Angabunga, such a situation being likely in most steep, swiftly flowing single thread channels.

Rapid changes in channel pattern also occur in the Adour in south-western France, whose longitudinal profile was described earlier. In this torrential regime, bankfull flow occurs 39 days of the year, with 12 to 14 alterations between underbank and overbank flow. Pebble bars in the channel shift rapidly. Upstream of the Arros confluence, small wavelength meanders reflect a relatively small discharge, steep channel slope and coarse sediment load. Between the junctions with the Arros and Mugron, alternations between meandering and braided channels correspond to greater discharge, moderate slope, and moderate grain size of bed material. The least stable section of the Adour, with its bed of well sorted debris, indicating continual adjustment of the channel to changing hydrological conditions, illustrates how the stream channel behaves as an open system, subject to modification as inputs of water and sediment fluctuate. Further downstream, where the channel gradient is less steep, the Adour has broad meanders, a bed load of sand and a higher discharge.

Over relatively short time spans human activity and, over long time spans, climatic and hydrologic changes alter channel patterns. Narrow, meandering, single-thread channels have been changed to wide, straight, braiding channels in New Zealand where removal of vegetation from hillslopes has caused an influx of coarse sediment into stream channels (Grant, 1965), and in some streams around Kuala Lumpur, Malaysia, where the clearing of land for urban development has so increased sediment yields that stream channels have been widened and straightened (Douglas, 1975).

BRAIDING CHANNELS

Braided streams occur in many humid regions, particularly in mountainous terrain. Eastern Australian rivers such as the Clarence and Macleay have braided reaches extending several kilometres downstream of their waterfall and gorge sectors incised into the New England Tableland. Rivers draining the Owen Stanley Ranges and the recently active Mt Victory and Mt Trafalgar volcanoes in eastern Papua have braided reaches where they flow across alluvial fans or the upper parts of alluvial plains at the foot of the mountains (Blake and Paijmans, 1973) (Plate 20). Both the eastern Australian and Papuan rivers change from braided to meandering channels as gradients become less further downstream. Streams with poorly sorted gravel beds develop braiding by the formation of low, linear, mid-channel bars at high flows and the dissection of bars at low flows. Flood flows thus build the bars at about peak discharge and then on the falling stage cut them into a series of small gravel areas divided by complex anabranches (Fig. 42). Frequent variations in discharge thus favour braiding as bedforms are constantly rearranged by successive flood flows, as in the Ramu River in the New Guinea Highlands.

The great rivers of the Indian sub-continent demonstrate the close relationships between discharge, sediment load and channel pattern particularly well. The Brahmaputra River channel migrates rapidly, shifting over 800 m in a single day. During low flows, small bed forms move about 120 m day^{-1}, but at high discharges, bed forms up to 15 m high travel downstream as quickly as 600 m day^{-1} (Coleman, 1968).

20 *Braided river in lowland tropical rain forest flowing north from the Torricelli Range, Papua New Guinea (CSIRO photo)*

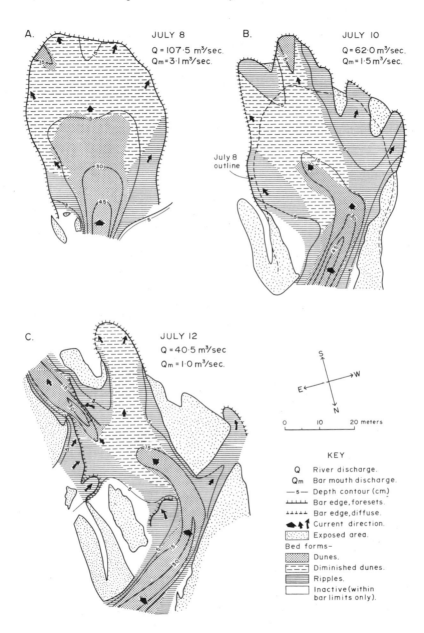

A.   JULY 8
Q = 107·5 m³/sec.
Qm = 3·1 m³/sec.

B.   JULY 10
Q = 62·0 m³/sec.
Qm = 1·5 m³/sec.

July 8
outline

C.   JULY 12
Q = 40·5 m³/sec
Qm = 1·0 m³/sec.

0   10   20 meters

KEY

Q   River discharge.
Qm   Bar mouth discharge.
—5—   Depth contour (cm).
⊥⊥⊥⊥   Bar edge, foresets.
⊥⊥⊥⊥   Bar edge, diffuse.
◄▲↑   Current direction.
        Exposed area.
Bed forms—
        Dunes.
        Diminished dunes.
        Ripples.
        Inactive (within
        bar limits only).

42   *Changes in bed forms in the Platte River, Nebraska at two-day intervals*
      during falling discharge (after Smith, 1971)

At any given periodic major channel bankfull discharge, braided channels occur on steeper slopes than meanders (Leopold and Wolman, 1957). Steeper slopes contribute to sediment transport and to bank erosion and are often associated with coarse heterogeneous materials. All these are conditions favouring braiding. Henderson (1963) argues that if transporting power is to be the criterion which distinguishes braided from meandering channels, then size of the bed material ought to be taken into account. When this is done, straight and meandering (or single-thread) channels generally conform to the relationship:

$$S = 0.6D^{1.4} Q_{BF}^{-0.44}$$

where

$$
\begin{aligned}
S &= \text{channel longitudinal slope} \\
D &= \text{bed material size} \\
Q_{BF} &= \text{bankfull discharge}
\end{aligned}
$$

For any given discharge and bed material size, meanders occur on the more gentle slopes, braiding on the steeper slopes.

CHANNEL PATTERNS AS A CONTINUUM

Although the processes of meandering and braiding have been described as distinct types of channel evolution, many rivers have braided reaches where they leave the hills, which change to meandering courses further downstream, while others change from meandering to braiding patterns. The Ganga River, for example, has fully developed meanders between Allahabad and Benares, but is braided downstream of the junction with the Gogra, Gandak and Sone Rivers (Chitale, 1970). Straight channels are not uncommon, the Damodar River near Durgapur barrage being an example of a wide, shallow, steep, straight channel carrying a heavy load of sediment with fairly uniform size bed material. To cope with these varied channel patterns, Chitale (1970) suggests that a logical classification would be that rivers have either a *single* or a *multiple channel*. Single channel streams can be meandering or straight, with a transition between the two, while there are two distinct types of multichannel streams, the interlaced multichannel stream known as a braided stream and the branching stream, such as a river building an alluvial fan or delta or developing distributary systems such as those of the Lachlan, Macquarie, Namoi and Gwydir in New South Wales.

Chitale's classification applies principally to rivers in alluvial plains. In mountainous and hilly terrain, channel patterns and characteristics are closely related to the relief and valley walls. Popov's fivefold classification (1964) into non-meandering embedded channels, channels with side-bars near steep valley-side slopes, limited meandering, free meandering, and braiding channels is probably the most universally applicable.

Non-meandering embedded channels. Where rivers flow in narrow valleys whose slopes form the channel limits, the channel has little freedom to move. Sediment is derived from transport down the channel and from the collapse of valley sides. Masses of bed load may be moved as large riffles or side bars, with small shifting islands being formed in exceptional cases. This type of channel is typical of rivers deeply incised into rugged terrain such as the Chandler River below Wollomombi Falls in northern New South Wales.

Channels with side bars near steep valley-side slopes. The valley floor is too constricted for the development of lateral deformations, but the rate of side bar shifting is at least as great as the rate of bank erosion. Eventually, the side bars may form continuous ribbons along

21   *The upper reaches of the Mahawali Ganga in the highlands of Sri Lanka near Carolina: a joint-controlled channel with boulder-size gravel side bars near steep valley-side slopes*

the channel, or may be broken up by oscillations of the talweg within the channel. The development of such side bars is particularly marked on rivers draining to the west coast from the highlands of Sri Lanka (Plate 21).

Limited meandering. Valley-side slopes are often not far enough apart for completely free meanders to develop. Meanders thus under-cut slopes, with channel sediment forming shoals or riffles on the insides of bends. As meaders shift downstream, the sediment accumulations also migrate. Such conditions occur in the Macleay River of eastern New South Wales just upstream of Bellbrook.

Free meandering. When the width of the floodplain greatly exceeds the width of the meander belt, loops and cutoffs develop frequently with consequent changes in channel length and channel slope. Flood-plain alluvium is deposited during overbank flows. Under conditions of free meandering, suspended and bed load transport acquires particularly complex forms as described earlier in the discussion of the Angabunga River of Papua.

Braiding. Considerable volumes of bed load and the slowing of bed load transport are associated with braiding. The shallow banks of debris in the river are highly mobile and channel form is unstable. In humid regions, braiding usually occurs as streams emerge from mountainous country, for example on the Bonua River south-east of Mt Suckling in eastern Papua (Blake and Paijmans, 1973).

CHANNEL PATTERN AND THE DRAINAGE NETWORK

The relationships between channel pattern and discharge discussed in this chapter suggest that as discharge is also related to basin area and drainage density, it may be possible to relate channel pattern to basin morphometry. The sinuosity of a stream, for example, tends to be generally related to stream order (Stall and Fok, 1968). In the Kaskaskia River, Illinois, sinuosity increases from about 1·4 to 2·0 as order increases from 4 to 6 (Stall and Yang, 1972). This increase of sinuosity downstream means that the stream is lengthening its flow distance as it flows further downstream. The slope of the stream is thus being reduced and consequently the unit stream power is decreasing in a downstream direction, in accord with the principle of the law of least time rate of energy expenditure.

Further exploration of the character of channel adjustments in different parts of the drainage basin will lead to better understanding of the functioning of the basin as a denudation system. In any

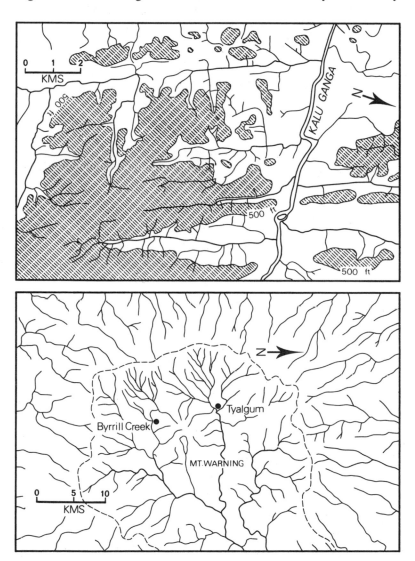

43 *Drainage patterns in the Kalu Ganga catchment, Sri Lanka (upper map) and the Mount Warning district, New South Wales (lower map)*

particular basin local geologic, climatic and biotic effects will cause deviations from general relationships based on the analysis of large numbers of gauging sites from widely separated rivers. An element of qualitative drainage network description thus becomes necessary.

Structure often controls the plan of a drainage network, with short gorge sections through resistant strata and long tributaries developed on weaker rocks. Twidale (1971) describes the theoretically possible pattern of stream development in folded sedimentary terrain, while Ollier (1969b) discusses radial and centripetal drainage networks in volcanic terrain. Examples of these and of a rectangular drainage pattern are shown on Figure 43. Often such a qualitative description aids understanding of how particular drainage networks have evolved and may depart from the laws of morphometry.

EXTREME EVENTS AND CHANNEL PATTERNS

The emphasis on trends towards equilibrium and regime conditions in this chapter needs to be qualified by a reminder that exceptional storms and floods can cause such vast channel changes that all the work of moderate and more usual flood flows can be obliterated in a single event. If the recent history of a river basin has included exceptional storms, the channels reflect the work of those flood events.

The impact of cyclones on tropical river systems may be great, causing floods such as those of January 1974 in the rivers of western Cape York Peninsula of Queensland which cut coastal sand bars and developed a complex series of new channels. Cyclone Brenda in 1968 caused the Dumbea River in New Caledonia to erode not only the banks facing the minor channel but the floodplain surface itself, creating new channels and redepositing the eroded material further downstream (Baltzer, 1972). Major floods, like those of December 1926 and January 1971 in West Malaysia, produce rapid channel changes, through the temporary damming of tributaries by debris and the development of cutoffs and new courses through old levee and terrace deposits. Human settlements have been destroyed, archaeologists finding the traces of floods in their investigations (Lineham, 1951). Temporary blockages of the channel suddenly give way, releasing great walls of water which roll down the valley sweeping away any unconsolidated material in their path. Where the valley widens, energy of downstream flow is dissipated laterally, the decrease in depth and velocity leading to the sudden deposition of ill-sorted coarse sand, pebble and boulder debris in large irregular mounds, one to two metres thick (Tricart, 1961). If such deposits result from the sudden breaching of a temporary blockage, they spread in a fan shape from the opening in the blockage.

Such flood deposits, laid down in a few hours or minutes, remain in position for many decades. For example, a large proportion of the deposits of the catastrophic flood of October 1940 in the Tech Valley of the eastern Pyrenees, France, remains perfectly recognisable and uncolonised by vegetation due to the infertility of the gravelly debris. The scour effects of such flows are also long lasting, the great floods of June 1957 in the Guil Valley of the southern French Alps, and those of 1926 and 1971 in the Sungai Kuantan of eastern Pahang, Malaysia, have left channels wider than those that would be developed by flows occurring between these exceptional flood events.

While in general most work of changing the landscape is done by events recurring as frequently as the mean annual flood (Wolman and Miller, 1960), individual drainage basins may have suffered from a particular extreme event, whose traces have not yet been effaced by the more regularly operating, but lesser magnitude geomorphic events. The long term trends towards regime conditions in channels is thus likely to be interrupted over time spans of hundreds or thousands of years (Tricart orders 7 and 6, Table 1) by catastrophic events which do more work than all the flows of several hundred or even thousand years before.

VIII

FLUVIAL
DEPOSITIONAL LANDFORMS

Meandering and braiding involve deposition and erosion of sediment, the shifting river and its occasional overbank flows gradually building up lateral deposits high enough to confine the majority of channel flows. The alluvial flat so built up is only inundated by flood flows and is known as the *floodplain*. Streams descending steep channels in the side of higher order major valleys deposit debris at the break of slope where stream gradient lessens, creating a fan shaped depositional cone, termed an *alluvial fan*. The landforms associated with these two features reflect the fluvial processes that create them, often providing valuable evidence of past changes in denudation and channel patterns.

The deposition and erosion of alluvial material by meandering and braided streams is an integral part of the process of floodplain formation. The channel bed, banks and floodplain may be regarded as a temporary store for sediment, into which material may be put, to be eroded away at other times. For most of the time, the flood plain is unaffected by the activity of the river, but during flood flows it may be the scene of large quantities of erosion and deposition. Occasional extreme events, such as the disastrous floods of 1949, 1950 and the early 1960s in the coastal streams of New South Wales (Warner, 1972), may erode, or re-deposit, vast quantities of deposition on the side of channel bends (point bars) (Plate 19) and deposition from overbank flows, discharge exceeding channel capacity must occur sufficiently often to maintain a supply of sediment to the floodplain flats. Thus the true active floodplain is said to be the zone inundated by floods recurring at least once in every one or two years (Fig. 44).

It was shown in Chapter VI that the active floodplain may be a relatively narrow inconspicuous depositional bench below the valley floor alluvial flat. Such a situation occurs in the Colo Valley near Sydney, New South Wales, where the valley floor, often 300 m wide

164

44 *Floodplain landforms along the Macleay River near Kempsey, New South Wales*

22 *Terraces in the valley of the Ramu River, Bismarck Range, Papua New Guinea (CSIRO photo)*

and 16 m above the low water level, is a former floodplain, abandoned by incision of the river (Hickin, 1970), the active floodplain being a narrow discontinuous strip about 8 m above the low water level of the river (Woodyer, 1970). Similar entrenchment has occurred in many valleys, leaving remnants of the former floodplain as stream *terraces* (Plate 22).

The morphology of floodplains

Although the relief of a floodplain is slight, every change of altitude affects the flood history of a particular locality. A difference in elevation of one metre may determine whether or not a point on a floodplain is subject to inundation and so to erosion or deposition. Many of the topographic features of floodplains may be differentiated by their elevation and the types of sediment of which they are composed. The condition of a meandering channel involves many phases of abandonment, cutoff and infilling of river bends.

Cutoff. As a meandering stream increases in sinuosity, there is an increasing probability that a high flow will break through the waist of a meander loop and create a cutoff. Studies of the Mississippi floodplain revealed two kinds of cutoff (Fisk, 1944, 1947), chute cutoff and neck cutoff. Chute cutoff occurs where a stream in a meander loop shortens its course by creating a new channel along some swale on the bar enclosed by the loop. Enlargement of the new channel and plugging of the old channel proceed gradually. Chute cutoffs are plugged mainly by bed load sediments which reduce the depth and width of the old channel. Eventually the suspended sediment deposited from overbank flows completes the infilling of the abandoned channel. Neck cutoff, the prime cause of the abandonment of meander loops, occurs late in the development of the loops, either through the carving of a new channel across the narrow neck of land between two loops, or by the cutting away of the outer bank of an upstream loop until it cuts into the adjacent downstream bend. Bed load rapidly blocks the ends of the abandoned channel, creating the familiar 'ox-bow' lake. Filling is completed by sediment deposited from overbank flows.

Avulsion. Meandering rivers occasionally abandon a part or the the whole of their meander course and create a new course at a lower level on the floodplain. This process of avulsion has occurred several times in the recent history of such streams as the Mississippi (Fisk,

1947), Meander in Turkey (Russell, 1954) and Angabunga (Speight, 1965a). While restricted to a meander belt, a stream builds up its bed and banks by deposition of bed load and suspended sediment, gradually increasing its elevation above the general level of the floodplain, forming an alluvial ridge (Fisk, 1947). The greater the height of the ridge above the floodplain, the greater the probability that local breaks through the meander banks will result in a permanent change of stream course away from the alluvial ridge, into a flanking basin where a new meander belt can be constructed. The Mississippi, for example, has recently built up five alluvial ridges in its floodplain along meander belts up to 80 km apart (Fisk, 1944).

Point bars. Lateral accretion of sediment on the inside of a river bed leads to the formation of a point bar. As flow increases, the width and depth of water rise, most of the increase of width involving the inundation of the point bar. The point bar is thus built up of material deposited from a succession of flows of varying depths. Generally, coarser material accumulates at the upstream edge of the point bar, which has a steeper slope facing upstream than on the downstream edge, which is usually lower and composed of finer particles. Although point bars at their lower limits are inundated before bankfull stage is reached, the upper parts of point bars merge into the floodplain and there is a gradual merging of material deposited by overbank flows with that derived from lateral accretion.

The development of point bars is thus probably a major factor in the evolution of floodplains, the lateral shifts of meanders leaving point bar deposits as new floodplain material. Thus lateral accretion in point bars, with small amounts of deposition from overbank flow, is the commonest means by which sediment is redeposited on the floodplains to compensate for that lost by bank erosion along the outer edges of meander bends.

Point bar evolution often involves the growth of a series of *meander scrolls*. These curving bars begin to form against the inner bank at the upstream end of the meander loop. Gradually the bar increases in height and extends downstream parallel to the inner bank but separated from it by a narrow, shallow swale. As the channel shifts laterally towards the outer bank, a new scroll may begin to grow. Through time a succession of abandoned meander scrolls may be left on the inner bank, with the swales between sometimes carrying long, narrow lakes.

On certain streams, point bars have a dune-like form, with a gentle slope facing upstream and a steep slope downstream. They are found in the tributaries of the lower Hawkesbury River in New South Wales, where the bed material is coarse non-cohesive sand (Hickin, 1969). These streams do not develop large meanders so that the point dune may be related to channel morphology, bed material and flow regimes.

Natural levees. The wedge-shaped ridges of sediment bordering stream channels, the natural levees, are highest close to the edge of the channel, commonly forming steep high banks. From the channel edge the levees slope gently away from the river into flood basins at some distance from the channel. Levee growth is largely responsible for the formation of alluvial ridges.

Overbank flows cause deposition on levees. The decrease in depth as water rises up the levee bank reduces velocity so that some of the sediment load can no longer be carried and is deposited on the banks. The coarsest debris is laid down close to the channel and finer material further down the levee away from the stream. Rates of deposition are greatest closest to the channel. Between overbank flows, surface runoff, raindrop splash and other processes of slope erosion attack the levee bank.

Crevassing and crevasse splays. Deposition beyond the main stream channel also occurs as a result of crevassing, the process by which water breaks out of the channel through isolated low sections or gaps in the levee banks, more commonly on the convex than the concave bank. Once a crevasse is initiated, the floodwaters deepen the new course and develop a system of distributary channels on the upper slopes of the levee. Sediment is deposited on the lower slopes, forming a tongue-shaped mass of sediment, or crevasse splay, which may extend into the flood basin and which results from deposition by branching and often braided streams (Plate 23). As the master channels of crevasses often carry water and sediment from well below the surface of the main river, crevasse splay deposits are usually coarser than the associated levee sediments.

Flood basins. In the lowest areas of floodplains are poorly drained, flat, relatively featureless areas, termed flood basins. These areas of little or no relief, adjacent to, or between, abandoned or active alluvial ridges, act as stilling basins in which fine suspended particles settle out of overbank flows after coarse suspended particles have been deposited on levees and crevasse splays.

23 *Meandering river showing dirty water spilling through a crevasse into a backswamp lake, Papua New Guinea (Photo by J. N. Jennings)*

Flood basins often have a network of small channels, partly inherited from older and more important drainage systems (Fisk, 1944). These channels perform two functions, carriage of water from an active meander belt into the floodplain during the growth of a flood, and the draining of water ponded in the basins back into the main river channel as the flood level drops. Probably the largest basin of this type is the Tonle Sap, or Great Lake, of Cambodia, which is filled by floodwaters of the Mekong during May to October and which drains out to the Mekong for the remainder of the year. As flood basins are generally limited in width by valley walls, they tend to be narrow and elongated, parallel to active streams.

Yazoo stream. Streams entering the floodplain cannot directly join the main river because the alluvial ridge of the main channel is at a higher elevation than the floodplain margin. The tributary therefore flows down valley, parallel to the river, until it enters the main channel at an appropriate elevation. Streams of this deferred junction type are termed Yazoo streams, the name being derived from the Yazoo River of the Mississippi River floodplain, even though the Yazoo River did not evolve in the manner described.

Construction of floodplains

Two broadly different groups of deposits are involved in flood-plain construction, lateral accretion and vertical accretion by over-bank deposition. The deposits of lateral floodplain accretion are formed along the sides of channels, where bed load material is moved by traction towards the convex banks of the channel. Normally such deposits of lateral accretion are later covered by finer material of vertical accretion, as the channel shifts further away by lateral bank cutting so that the convex, slip-off slope on the inside of the bend is inundated less frequently and at lower velocities. Vertical accretion, by building levees, crevasse splays and flood basin deposits, is mainly the result of the deposition of material carried as suspended sediment. Channel fills and lateral accretion deposits are largely bed load material. However, distinction between the two forms of bed material deposition is not easy.

Some floodplains lack vertical accretion features, or have them only to a limited degree, but lateral accretion deposits are common to all floodplains. Whether significant vertical accretion occurs or

24   *Floodplain and two terraces of the Pahau River, South Island, New Zealand. The terraces probably result from regional uplift. Well developed meander cusps in the upper terrace. (Photo by J. N. Jennings)*

not probably depends on internal factors, inherent in the stream regime, and on others external to the stream. The important internal factors are probably the size of suspended sediment particles, the general calibre of the bed load, the rate of channel migration, and the speed of overbank flows. The external factors are changes in the level of the water body into which the stream flows and changes of land level due to compaction of alluvial sediments, isostatic adjustment or tectonic movements (Plate 24).

Catchment conditions greatly affect the behaviour of a river on its floodplain. The Sepik River floodplain in New Guinea is aggrading steadily, the river channel having been displaced towards the northern edge of the plain, probably as a result of the greater volumes of sediment brought down and deposited by tributaries entering the southern edge of the floodplain from the high mountains of central New Guinea (Reiner and Mabbutt, 1968).

Floodplains represent an adjustment to the discharge and sediment characteristics of a river. As indicated in the previous chapter, reduction of discharge or increase of sediment load change the river pattern and its relationship to the floodplain. Adjustments may be completely wrecked by a single storm event, hurricane Camille depositing a layer of debris from 0·06 to 1·5 m thick over the floodplains of Davis Creek and Rucker Run in Virginia (Williams and Guy, 1973).

ALLUVIAL FANS

Basically, fans are formed by deposition of alluvial material beyond the limits of mountain valleys as the result of changes in the hydraulic geometry of flow after the stream leaves the confines of the trunk stream channel. When a stream reaches the change of slope on leaving the mountain valley, it widens, but decreases in depth and velocity, so depositing sediment (Cooke and Warren, 1973). Many fans have the same gradient as the river bed in the bedrock above them; the building of the fan has eliminated the former break of slope which initiated their growth (Plate 25).

Among the greatest alluvial fans are those formed by the Himalayan rivers entering the Indo-Gangetic Plain. Rivers flowing over this piedmont alluvial plain are characteristically braided and frequently change course (de Martonne, 1951). Within the Indo-Gangetic plain may be discerned two major alternating relief forms,

25 *Active alluvial fan on north side of the Markham fault trough and at foot of the Sawtooth Range, Papua New Guinea (CSIRO photo)*

the great alluvial fan, spreading out from the gorge exits of all the great Himalayan rivers except the Gogra, and the intervening slopes which may be called the inter-fan areas (Geddes, 1960). The fans, triangular or rather segmental in plan and convex in form, have their apices at the gorge mouth from which their axes run to their bases towards the Deccan Plateau edge, or the sea; the inter-fan areas taper from the Himalayas and are slightly concave at the edges of their even slopes. Fans with gradients of over 1 in 5000, and the inter-fans, with somewhat less, merge, where space allows, in a fan-foreset plain with a gradient of 1 in 10,000 or less, which continues to the delta and sea.

The plain of the north Bihar between the Himalayas and the Ganga consists of the three fans of the Gandah, the Kosi and, in part, the composite Mahananda-Tista fan, with the two inter-fan areas between. These three fans exhibit a radial pattern of distri-butaries. Those of the Kosi have moved westward for the last 150 years, their floods and deposition of micaceous sands causing co-lossal damage, thus earning the river the epithet of 'Bihar's sorrow' (Geddes, 1960). The westward movement over a total of 110 km has taken place sporadically, the river shifting as much as 19 km in a single year (Leopold, Wolman and Miller, 1964). Even in a normal season the Kosi's sands may raise the level of the floodplain by 30 cm. Following the Bihar-Nepal earthquake, which produced land-slides on the foothills, even the margins of the flood zone had been raised by over 30 cm.

The minor relief pattern of each alluvial fan is complex, reflecting the scour of channel beds and lateral bank erosion which occurs when flow is contained within distributary channels after floods. New deposits are laid down in radial bands, sometimes terminating in lobe-shaped zones, the areas of deposition shifting with the changes in distributary position. Alluvial fan sediments become finer with increasing distance from fan apex (Allen, 1970), the Kosi having boulders in the foothills gorge, pebbles at the apex of the fan, and then sands, which become finer downstream. In an extreme flood a big distributary, flowing down the fan at great velocity, swings to either side, spilling over the channel banks in many places. Where the overbank flow is checked, the sands drop, forming 'spill-banks' or natural levees, beyond which the size of deposits decreases until fine clay only is deposited in flood basin depressions between distributaries (Geddes, 1960). Thus on alluvial fans as large as those of the Indo-Gangetic Plain, an alluvial topography analogous to that of the floodplain of a meandering stream is developed about every distributary.

In many mountainous areas of Papua, large, coalescent alluvial fans, 'alluvial aprons', lie along the foot of steep ridges such as those drained by the Mambare River north-west of Kokoda (Blake and Paijmans, 1973). These alluvial fans develop in the same way as do those of the Indo-Gangetic Plain, because the debris shed by the high-relief areas cannot be carried off immediately on the gradient of the lowland streams, and the debris is stored in the piedmont zone until it is reduced by weathering and erosion to finer sizes. Successive layers of gravel, sand, silt and clay deposited on the fans by the streams disgorging from the mountain ridge gradually extend the apex of the fan further upstream, replacing the original break of slope with a less marked break of slope further upstream and a gentler slope down to the major river which cuts the base of the fan. The area of alluvial fans is proportional to the area of the drainage basin in the mountains and the difference in rock resistance and relief between the mountain and lowland areas.

Alluvial fans change character rapidly as the front of the fan is cut by the main stream of the valley and as distributaries cut new channels across the face of the fan. Eventually dissection and lateral erosion of the fan may convert spreads of piedmont alluvium into a complex of gravelly floodplains, terraces and dissected terraces.

Among Australian examples of alluvial fans are those created by streams cascading down the eastern slopes of the Bartle Frere-Bellenden Ker massif to debouch onto the Mulgrave-Russell plain. These fans are restricted in extent because weathering of the debris and its erosion by intense cyclonic rainfalls removes fan materials rapidly. Alluvial fans are found on the edges of many temperate coastal plains in eastern Australia, as at Camden Haven, New South Wales (Fig. 45). Well developed fans bordering the Murray Valley near Albury have sometimes been truncated by the lateral migration of the main stream (Hills, 1940).

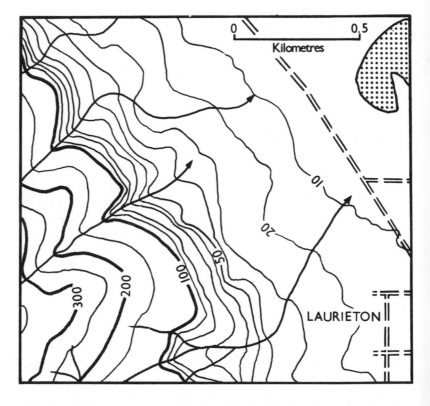

45 *Alluvial fan on the eastern flank of North Brother Mountain, Camden Haven, New South Wales*

Alluvial cones

As the French term *cone de déjection* is widely used for all kinds of alluvial fan formations, some writers refer to the alluvial fan landforms as alluvial cones. However, Hills (1940) and Thornbury (1954) both point out that an alluvial cone, although similar to an alluvial fan, is steeper and smaller, being laid down by small streams flowing down escarpments and fault scarps such as those on the Lake George escarpment near Canberra, Australia (Wasson, 1974). Such small streams often flow only after heavy rain, when the whole surface of the cone may be flooded with a shallow sheet of water, termed a sheet flood. Alluvial cones, therefore, lie between talus cones, formed without the action of running water, and alluvial fans in sedimentological characteristics.

TERRACES

In the discussion of floodplains it was noted that many stream channels are becoming entrenched with the creation of new floodplains below alluvial valley floor flats. Remnants of earlier floodplains, now abandoned by the river, are termed alluvial terraces. Prominent terraces of this type occur on all the major rivers of eastern New South Wales that emerge from gorges on to narrow coastal plains. Five distinct terraces have been identified in the Macleay Valley upstream of Kempsey (Walker, 1970); fragments of three have been mapped in the Manning Valley upstream of Taree (Davis, 1965); two major and additional minor terraces occur in the Hunter Valley (Galloway, 1963); and similar sets of terraces are found in the Shoalhaven Valley upstream of Nowra.

If a river cuts its channel down as part of the adjustment of gradient towards a smoother profile, it will eventually cease to flood a segment of the floodplain regularly. Abandonment of portions of the floodplain would occur irregularly, giving rise to uneven terrace remnants on different sides of the valley. Lateral migration of the channel assists such evolution of a series of 'unpaired' terraces.

More commonly pairs of terraces on either side of the channel are found. They arise because the river has started to behave differently, either because its outlet has been lowered or because conditions in the catchment area have altered. The New South Wales terraces are thought to have developed as a result of changes in river regime or sediment yield (Warner, 1972), possibly caused by climatic change.

A

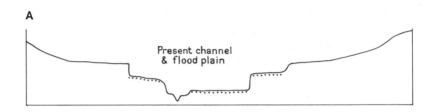

B

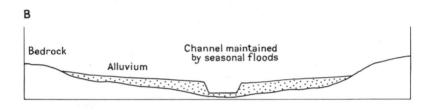

C

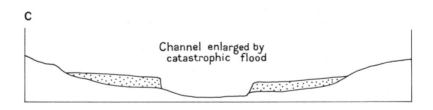

D

46   *Development of terraces by lateral migration of a stream (A) and by major flood erosion (B, C, D)*

As catastrophic floods may drastically alter floodplains, they may be responsible for some valley-side benches. Hack (1965) considered that while climatic changes could explain some of the terraces in the Shenandoah Valley, other terrace features arise from catastrophic floods, which, although rare, are part of the present climatic regime. Before the flood the river may have had a fairly narrow channel through a wide stretch of alluvium. During the flood the river could cut a much wider, and locally deeper, channel. Subsequently the river would deposit new alluvium within the flood channel, so creating a new, inset floodplain below the pre-flood deposits (Fig. 46).

Identification of terraces

To identify terraces, morphological characteristics have to be compared with the lithology, granulometry and fabric of the terrace deposits together with the development of soils on the alluvial surfaces. Other marker beds aiding recognition of terraces are provided by the fossil mollusca and petrographically distinctive volcanic ash to be found in some terrace deposits. Because ash beds are usually laid down quickly and over a wide area, they are particularly useful for the correlation of depositional surfaces. Even if terrace remnants are scattered over a wide area, the identification of the lowermost ash bed in any terrace remnant enables it to be correlated with any other fragment of a terrace having the same ash in the lowermost ash deposit. Thus in New Zealand, the terraces of many valleys in the Bay of Plenty area have been easily dated and mapped (Pullar, 1965; Pullar, Pain and Johns, 1967).

Where terraces do not contain distinguishing fossils or marker beds, sedimentological analyses aid the correlation of terrace fragments. In studying the Quaternary evolution of north-western Spain, Raynal and Nonn (1968) examined the sedimentary structures, weathering characteristics, morphometry and granulometry of pebbles and sands, clay mineralogy and lime content of a variety of alluvial deposits. The contrasts in sedimentological characteristics of deposits of different age enabled them to identify and correlate terrace remnants.

The characteristics of soils on terrace remnants are most valuable distinguishing criteria. Soils on the five terraces along the Macleay River, New South Wales, show a progressive increase in the degree of soil development with age (Walker, 1970). The older profiles have greater particle-size differentiation and more development of illuviated clay (Brewer and Walker, 1969). Following Butler (1959), the lowest terrace remnant in the Macleay Valley was designated K_1, and the successively older terraces, K_2, K_3, K_4, and so on. In general, the frequency of argillaceous lithorelicts increased with depth in each terrace soil profile, but decreased from the K_1 to K_5 profile. The ratio of argillaceous to quartzitic fragments changes in the same way: biotite is recognisable throughout the K_1 profile, in the lower part of the K_2 profile, and not at all in the older profiles. These data indicate increasing weathering with increasing age from the K_1 to the K_4 profile (Brewer and Walker, 1969).

Positive indications of the actual age of terrace remnants may be gained from archaeological evidence and radiometric dating. The difficulties that arise from the washing of artifacts into deposits or the contamination of organic remains by recent material make the establishment of accurate alluvial histories hazardous. Great care must be taken to avoid unwarranted inferences about rates of terrace formation, erosion and dissection (Vita-Finzi, 1974).

Non-depositional terraces

Not all terraces are formed by the deposition and dissection of valley floor alluvial fills. In some valleys, stripped structural surfaces occur where erosion has stripped less resistant rocks from flat-lying resistant rocks, such as sandstone or limestone. In New South Wales rivers flowing through the nearly horizontal Hawkesbury Sandstone frequently have flat benches paralleling the river.

As a river shifts across the floor of a confined valley it undercuts the valley walls and develops a rock-cut floor. If the river then cuts into the valley floor, the rock-cut remnants of the former floor will become benches above the level of the present channel. These lateral planation platforms generally match in height on opposite sides of the valley, unless the surface is raised locally by slopewash from bordering areas. Often such rock-cut benches carry thin coverings of gravels associated with former floodplain levels.

These rock-cut benches or terraces may later become buried by alluvial floodplain material. Alluvial terraces may give way to rock-cut terraces along a single stream if resistant rocks outcrop along the channel. As the topographic form of the two types of terrace in the one valley is similar their origins could only be discovered by close examination of terrace stratigraphy. Mass movements in mountainous humid terrain may create valley fills, which are cut into terrace-like forms by rivers. Such events are part of the present denudation regime and the forms derived from them should not be confused with true terraces.

Origins of river terraces

Terrace formation by river entrenchment into a valley floor deposit may result from:

a eustatic changes in sea level;
b effects of glaciation on load and volume of streams heading in glaciated areas;
c effects of climatic change on the regime of rivers rising in non-glaciated areas;
d crustal warping;
e combinations of these influences (Smith, 1949).

Eustatic sea level change, the upward and downward movement of sea level, is world wide because the oceans are interconnected (Bird, 1968: 31). The withdrawal of large amounts of water from the oceans during the enlargement of continental ice sheets lowered sea level and caused streams to adjust their gradients, thereby cutting down through valley floor deposits to a new base-level. The rise of sea level, on the other hand, caused aggradation of channels and the burying of terrace remnants.

Tectonic uplift causes dissection of valley floor deposits, whereas downward movement provokes aggradation. Differential uplift in the Alpine orogenic zones, such as New Guinea and New Zealand, has produced widely separated terrace levels with differing gradients, the older terraces having been lifted more than the younger ones. Uplift across a major river may warp the relative position of terraces, raising sections of them higher above the present channel than unaffected segments upstream or downstream of the uplifted area. In Hungary the relative heights of Quaternary terraces range from a Holocene level of 2-3 m to the oldest at 65-90 m, but in the gorge of Visegrad on the Danube, Pleistocene uplift has raised terraces to 200 m above the present river bed (Ronai, 1965). The uplift of the Carpathians and downward movement of both the Pannonian basin and the Wallach depression of Romania have caused early Pleistocene terraces to be at widely differing altitudes.

The terraces of many northern European rivers, the Danube, Rhine, Seine and Thames, are glaciofluvial terraces, those of the

Thames Valley near Standlake producing a notably asymmetrical valley, with the terraces fed by north bank tributaries such as the Windrush having pushed the major stream close to the southern side of the valley. In eastern Austria, as in nearly all humid mid-latitude environments, periglacial frost weathering and solifluction supplied large quantities of debris and gravel from which thick fluvial terraces resulted (Fink, 1965). Not only did larger streams connected with glacier tongues form terraces, but so did small streams affected by periglacial action. For example, the Pershling, a small river in the eastern part of the northern foothills of the Alps, produced gravel terraces of the same thickness as the River Traisen, a much more important stream which was fed by a local glacier (Fink, 1961).

The impact of climatic change on river behaviour depends on the direction of the change and the relative importance of discharge and sediment load changes. A sequence of paired terraces in central Texas is believed to have evolved by the downcutting of the river under humid conditions, and the widening of the channel with slight aggradation during drier periods when surface runoff and sediment yield were higher. Such behaviour accords with the transitions from meandering to braided condition in streams affected by higher sediment yields and flood peaks as a result of human activity. In many areas terraces have evolved through combinations of eustatic change, climatic change and tectonic influences, as in the South Island of New Zealand. Great care is therefore needed in the analysis of river terrace evolution.

LAKES

Lakes are waterbodies due either to a barrier of some sort being formed across a river system or to the formation by various agencies of closed depressions in the relief. Lakes may be part of a river system, with inlets and an outlet, as in Lake Kutubu in the central highlands of New Guinea, or basins with no outlets, such as Lake George on the Southern Tablelands of New South Wales, or basins with virtually no inflow or outflow such as the crater lakes of Eacham and Barrine in north Queensland. In humid climates, absence of an outlet to a lake is unusual, generally arising either because the lakes have been recently formed, or because an underground outlet is available as at Lake Tebera in the same part of New Guinea as Lake Kutubu.

Lakes forming on new land areas, and those resulting from fluvial action, landslides and organic activity, are of particular relevance to the study of humid landforms. Although other classes of lake are found in present-day humid areas, their origins are discussed most properly in the context of other types of landform evolution, be they coastal (Bird, 1968), volcanic (Ollier, 1969b), structural (Twidale, 1971), glacial (Davies, 1969), karstic (Jennings, 1971) or desert (Mabbutt, 1977).

A lake may provide valuable clues to the analysis of the more recent history of the earth's surface. By skilful mapping of the landforms and by careful examination of the deposits, former lakes have been detected in many parts of the world. By studying the processes that are in operation today it can be seen that even the lakes that now exist will one day disappear. However, their fluctuating levels and capacities tell much of what is happening in the catchment as a whole.

Lakes on new land areas

The emergence of land from the sea results in the preservation in the landforms of the depressions which may have existed on the sea floor. Such depressions frequently become lakes, as in Florida where some of the low altitude, shallow lakes on a surface relatively recently emerged from the sea are of this origin (Harding, 1942). Emergence lakes may be short-lived, as rapid erosion of their outlets may drain them dry or erosion in their catchment areas may result in rapid filling of the lake with sediment.

Lakes resulting from fluvial action

Floodplain evolution may involve many types of lake formation. In addition to oxbow lakes formed by meander cutoffs and lakes in other river channel remnants and swales between meander scrolls, lakes form when levees block the outlets of tributary streams. Along the Pahang River in West Malaysia great quantities of sediment deposited on levees during floods block small lowland tributaries, creating lakes such as the Tasek Bera and Tasek Chini. Wilhelmy (1958) describes such lakes as *Dammuferseen* (embankment lakes). He discusses lake basins enclosed within a cutoff meander loop (*Umlaufseen*), but he refers also to lakes in depressions between two active river arms (*Inselseen*), to lakes due to the damming on

26 *Floodplain of the Aramaia River, Papua New Guinea, with backplain
lakes (Rückstauseen) in the background (CSIRO photo)*

floodplains of water brought down by tributaries (*Rückstauseen*)
and to those lying simply between the upland and the river without
such tributary discharge (Plate 26).

Lake Pepin, on the Mississippi River, formed by the deposit of
alluvial material brought in by the Chippewa River, is an example of
a *Rückstausee*. This lake in turn has been partly filled by silt carried
by the Mississippi River. The Mississippi has formed Lake St Croix,
a *Dammufersee*, by depositing material across the mouth of the St
Croix River. Tulare Lake, in California, was formed in the San
Joaquin Valley by an alluvial ridge deposited across the valley by
the Kings River (Harding, 1942).

A group of floodplain lakes with no tributary inflow occurs in the
Ka Valley of New Guinea. Although there are no topographic levees
in this valley, silt and clay deposits fringe the river with peats stret-
ching from the edge of these riverine deposits to the steeper ground
at the edge of the floodplain. Peat has formed at approximately the
same rate as silt and clay is deposited by the river, save in a few
localities where the fen developed more slowly and the river
occasionally spilled over on to the plain. As peat continued to
develop in surrounding areas, the lakes gradually became persistent
features of the floodplain, often without a channel linking them directly to the river (Jennings, 1963). Here there is a delicate balance
between fluvial action and peat development. Should flood mag-

nitudes or silt loads carried by the river change, then the stability of the Ka Valley lakes may be endangered.

Lakes resulting from landslides

Landslides and mass movement of material down valley-sides may create lakes, such as Lake Tarli Karng near Mount Wellington, the deepest lake in Victoria (Bayly and Williams, 1973), by blocking rivers. Earthquake initiated landslides often create lakes, as in the upper Madison west of the Yellowstone National Park, where the August 1959 earthquake triggered off a slide of 35 million m^3 of broken rock right across the river valley. The water of the river began to accumulate rapidly behind the new barrier and to prevent the sudden creation and enlargement of an overflow channel the U.S. Corps of Engineers bulldozed an artificial spillway which allowed surplus water to escape from the lake at a controlled rate (Shelton, 1966).

Lakes resulting from organic activity

Organic activity may occasionally give rise to lakes. In forested areas rivers are sometimes blocked by floating trees washed down during floods. The development of a log-jam forms a dam which holds back a lake. Eventually the dam bursts, but in the 1926 flood in Malaysia, this type of temporary lake developed new outlets which eventually became new routes for river channels. Such log-jams may produce changes in the drainage network. More long-lasting lakes of this type are produced by the dams constructed by beavers in the rivers of north-western Canada. The Ka Valley lakes described above illustrate the role of peat in the creation of lakes. Generally, however, lakes resulting from organic activity tend to be small.

Geomorphic processes in lakes

Any lake of some extent will develop waves which will lead to geomorphic action on the lake shore and the formation of beaches. Tides and seiches occur in lakes but they have little or no geomorphic significance (Hayford, 1922; Horton, 1927; Harding, 1942).

Wave and wind action in lakes affect lake shores in a similar manner to the action of the sea on the coastline. In large lakes many features of marine coastal processes occur. Lake Balaton (central

Transdanubia, Hungary) exhibits alarming rates of beach erosion and silting. On one portion of the shore, a section 110 m wide was eroded away between 1896 and 1943. The eroded fine sediment is carried towards the heavily silting Keszthely Bay by north-easterly currents (Hamvas, 1967).

Lakes and hydrology

Lake levels vary in response to hydrometeorologic events in their catchments and fluctuations in the rate of evaporation over their own surfaces. Lakes represent a natural form of storage, tending to reduce peak stream flows. They may also increase the low flow of the stream unless the evaporation from the lake exceeds the effect of the pondage. These equalising effects of pondage depend on the constrictions of the outlet as well as on variations in inflow. Lakes that trap sediments may give rise to changes in water quality. Lake levels tend to fluctuate not only in response to input and out-put factors but also relation to climatic changes, tectonic events and human interference. Examination of lake shorelines may reveal evidence of former higher lake levels in the form of wave-cut notches and raised beach deposits. Temple (1964) finding three raised beaches up to 20 m above the present level of Lake Victoria in Africa. Above these beaches are lacustrine remnants whose altitude has been affected by tectonic warping. The three underformed raised beaches belong to a 20-25,000 year period of tectonic stability when the lake level was lowered by the downcutting of the Nile outlet (Temple, 1969; Bishop and Trendall, 1967).

In the shorter term, lake level fluctuations of considerable magnitude have occurred over the last seventy years in response to relatively slight changes in climatic conditions. Rise or fall of the level of Lake Victoria is almost exactly proportional to the excess or deficit of rainfall compared with the average (Temple, 1964). Mörth (1965) has found a high correlation ($r = +0.96$) between month to month changes in the level of Lake Victoria in the period 1938-64 and rainfall over the catchment area, which, including the lake, is some 196,000 km^2 in extent. All the great lakes of Africa display similar variations in level, which are ultimately traceable to runs of wetter or drier years (Lamb, 1966).

However, as lake morphology tends to change rapidly, particularly at and around the outlet, the physical controls of lake level are apt to alter through time. Lake Tanganyika shows bigger fluctuations in lake level than Lake Victoria as a result of the silting up of the outlet bar (allowing the lake to rise somewhat) in dry periods and of breaching and removal of it when wetter years pile up too much water (Lamb, 1966).

The elimination of lakes

All lakes are destined to have a more or less limited existence. Outlet erosion can result in the complete drainage of lakes. Rivers fill them with sediment at rates which depend on the catchment area and morphology of the lake. Where catchments are small, as in the case of crater lakes, infilling with sediment is insignificant, but where the catchment area is large, with conditions favouring sediment production, infilling may be rapid.

Sediment brought into a lake by a stream is deposited as the flow velocity falls where the stream looses all gradient. The coarsest parts of the load are dropped closest to the mouth of the inflowing stream and finer particles are carried out further into the lake. The finer parts of the suspended sediment load and colloidal particles are held in the water until flocculated into sufficiently large particles to sink under the influence of gravity. Deltaic sedimentation of this type can fill lakes quite rapidly, many glacial lakes being filled within 2000 years or so.

Shallow lakes with small fluvial sediment inputs are filled mainly by vegetation. Calcareous and organic muds usually come first, then reed-swamp and fen peats. Moss peat bogs become established as domed floating mats above lake levels, but decayed organic matter and silt accumulate on the floor beneath them as gel mud, *dy* ('sedimentary peat'), in moderately acidic waters, or in well aerated water as a grey-black or brownish sediment, known as *gyttja*. Plants with roots colonise the margins of the lake out to about 1 m depth. As the roots impede wave action and plants supply abundant organic matter in the near shore zone, a net accumulation of organic matter progressively narrows the open water surface, and is followed by colonisation of the peaty shores by swamp plants.

At the same time a floating raft of sphagnum builds up and expands until the lake surface is covered by this raft. As sphagnum can only live in acidic conditions, the colonisation of the lake can only proceed if the input of organic matter is sufficient to create the necessary acidity. Complete colonisation sees the death of the sphagnum and the colonisation of the lake surface by trees. However, the process is not continuous, major floods scouring out the peat and burying vegetation beneath inputs of sediment.

DELTAS

Barrell (1912) defined a delta as 'a deposit partly subaerial built by a river into or against a body of permanent water' adding the important comment that 'a delta consists of a combination of terrestrial and marine, or at least lacustrine strata, and differs from other modes of sedimentation in this respect'. In the discussion which follows attention is focused on lacustrine deltas. For coastal deltas the reader is referred to the work by Bird (1968).

Delta morphology

A delta frequently comprises three sedimentation zones, topset, foreset and bottomset beds (Fig. 47). The surface of the topset beds may be divided into a subaerial and subaqueous deltaic plain, each of which may be further subdivided on the basis of sediment type and surface morphological features.

A delta is normally made up of numerous active and abandoned river channels, more or less marked levee ridges, and shallow water levee basins of various types (Plate 27). Where unconstricted the delta will have a lobate form, but digitate levees may extend out into the lake if the velocity of the inflowing water is high, for example in Wandandian Creek, N.S.W. (Fig. 48). Many lake basins are confined by steep mountain walls, such as Lake Laiture in northern Sweden (Axelsson, 1967) and Upper Arrow Lake, British Columbia (Fulton and Pullen, 1969) where the deltas occupy the full width of the head of the lake. Upper Arrow Lake occupies a section of the valley of the Columbia River. The only apparent difference between the part of the valley occupied by river and that occupied by lake is that the river-occupied section contains a fill of sediment, whereas the lake-occupied section is filled with water.

Deltaic deposition

Grove Karl Gilbert's study of the topographic features of lake shores led him to recognise that the formation of a delta depends almost entirely on the following laws: 'The capacity and competence of a stream varies with the velocity (the capacity is the total load of a given calibre which the stream can carry; the competence is the maximum size of the particles which the stream can move on the bottom). Thus if the velocity diminishes quickly at the mouth, deposition will occur on the bottom of the channel'. Such bottom de-

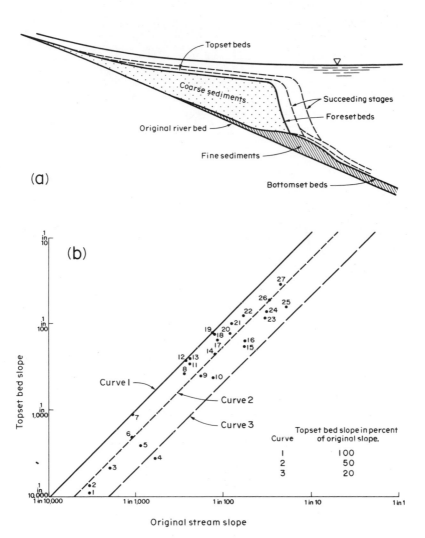

47 *(a) aspects of lacustrine delta morphology*
(b) relationship between topset bed slope and original stream slope for various lacustrine and reservoir deltas in the U.S.A. (After Borland, 1971.)

27 *Deltas of small braiding rivers from the Saruwaged Range on the north coast of Papua New Guinea (Crown Copyright reserved. By courtesy of the Director of National Mapping, Commonwealth of Australia)*

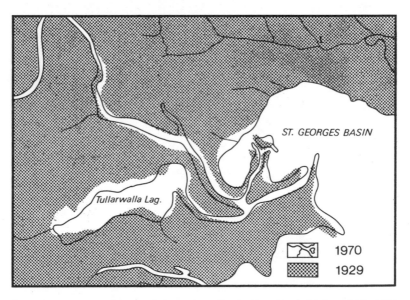

ST. GEORGES BASIN

Tullarwalla Lag.

1970
1929

48 *Changes in the delta of Wandandian Creek, New South Wales, 1929–1970*

posits form an obstacle to the current and promote continuous deposition upstream until the profile of the river acquires a continuous slope upstream from the extending mouth to the original course above the delta. The slope must be sufficient for the velocity of the stream to be adequate for its load.

Not all deltas show these precise relationships, but any stream will tend to maintain a balance between its slope and the work it has to do. Since deposition starts at some distance from the mouth the decreased load may be transported over a gentler gradient; hence the profile becomes slightly concave. At the outer end of the deposit, there is often a sharp double break of slope: the slope of the surface of the stream changes to the horizontal surface of the sea or lake— any tide will therefore be a disturbing factor—and the slope of the stream bottom changes to the much steeper one of the delta front.

As the current is swifter at the centre than at the side, natural banks or levees are built up on each side. However, these banks are never as high as the level of the greatest floods throughout their length. Therefore, accidental changes occur: the banks are breached, the breach is rapidly eroded, and the new channel is cut deeper than the old one, which is quite likely to be abandoned. This is all the more likely if the old course has a very gentle slope, while the new channel presents a shorter and therefore steeper route to the sea or lake. The position of the mouth varies and the stream tends to build up a semicircular delta, comparable with an alluvial fan in form but much flatter.

To illustrate the complex processes of lake delta evolution, the Kvikkjokk delta in Swedish Lapland, a joint delta of two confluent rivers at the west end of Lake Saggat (Dahlskog, 1969), may be cited.

The delta now has an area of 3·6 km² of which 2·07 km² are above normal summer low water level and 1·57 km² consist of channels and delta lakes. According to measurements on maps and recent aerial photographs, the growth of the delta in the last 80 years has been about 500 m² annually. The delta is flooded at least once a year. At the flood level, taken to be 0·45 m above low water at least, the riverward swales between scrolls on levees are flooded, the sedge fens and at least low willow thickets around delta lakes with channel connections to the rivers are inundated, all uncolonised sand bars and banks and most of the pioneer vegetation in the downstream parts of the delta are under water, and most levees in the downstream part of the delta, except those that have partly developed by aeolian sedimentation, are at least partly flooded.

The sediment load transported annually to the delta varies between 2000 and 10,000 metric tons but over 50 per cent of the year's total transport—more than 6000 tons—may flood the delta in a single day. The deposition of this sediment reaches maximum values on disturbed sites with strong and sudden shifts of currents: erosion and deposition alternate with short intervals in both time and space, and here sedimentation of about 1 m each flood occurs. Sites in more quiet and steady development where deposition is normal and erosion is very rare (the riverward slope of the outermost scroll on a levee ridge or meander bend, a growing bar or bank in a quiet water area with no strong or sudden flood currents) generally have a maximum sedimentation of 10 to 15 mm, rarely up to 100 mm each flood. The factor which more than any other determines the sedimentation in a locality is the distance to an active channel. Localities far from active channels receive no sedimentation at all, even though they are inundated for long periods and sediment transport is very high. The inside slopes of levees, bars and scrolls are flooded suddenly when the crests are overtopped, and on such slopes sedimentation is a function of the distance from the crest and generally decreases as height decreases. The outside slopes are flooded gradually, and only in such localities does sedimentation increase with decreasing elevation.

In most deltas there is an intricate relationship between vegetation and delta development, and a succession of vegetation development can be recognised. Vegetation is a response to the net effect over a number of years at each locality of several factors of varying importance: sedimentation on the other hand is a direct result of the factors prevailing during a particular flood. Vegetation has a history, sedimentation has not.

Nevertheless the pioneer vegetation communities are dependent on sedimentation, in so far as they either demand it or are able to tolerate it when other communities cannot. Variations in bottom layer and in frequency combinations give a wide range of pioneer communities, although only few vascular species occur regularly in pioneer vegetation. In the Kvikkjokk Delta silt and finer material are colonised by the aquatic *Isoetes* and the grass *Alopecurus aequalis* var. *natans*.

This usually short-lived community is transformed by the invasion of the grass, *Deschampsia caespitosa,* and the sedge, *Carex aquatilis.* At higher levels above low water deposition from bed load leaves coarser material on which there is a close adjustment between plant communities and varying environmental conditions. As the elevation increases and sedimentation begins to decrease, young willows and ground cover vegetation invade, with a birch environment developing when the ground becomes less subject to flood.

Mechanics of delta development

Laboratory flume experiments on the effect of varying sediment loads on delta shape (Chang, Simons and Brooks, 1967) produced deltas of half elliptical form and suggested that a shape factor would assist the understanding of delta growth:

$$\text{shape factor} \quad \frac{\text{Width of the elliptical delta}}{\text{Length of the elliptical delta}} = \frac{W}{L}$$

The experiments showed that the shape factor increased as the channel became aggraded. Thus with channel deposition the delta becomes wider. The delta was lengthened during channel degradation, producing a decrease in the shape factor.

Although these experiments indicated the type of processes and landform evolution which may occur in deltas, in practice deltas often show the effects of varying lake levels; of periods of advance and retreat of the delta front; of changes in the relative importance of distributaries and of major floods on patterns of sedimentation. In an attempt to see whether overall models of delta development could be applied to the major deltas of the world, Wright and Coleman (1973) found so much natural delta landscape variety that they concluded no general predictive model could provide reliable forecasts of delta morphology.

GEOMORPHIC HISTORY FROM FLUVIAL DEPOSITS

Floodplain deposits, alluvial terraces, lakes and deltas, all contain sedimentary evidence of past processes. While interpretation of these sediments has its difficulties, they provide vital evidence for historical geomorphology, the explanation of how the landscape has evolved through time.

On the shortest time scale, every major rise in a river will alter the

channel bed and thus each layer of sediment is a record of a particular flood flow. Close study of sand-sized bed material enables
the various flow stages, ripples, dunes and flat bed, to be recognised,
Moss (1972) suggesting that the fine to coarse ripple bed stage transition is temperature-sensitive and could be used as a basis for es-
timating water temperature at the time the deposit was formed.

On the longer time scale of hundreds, thousands or even millions
of years, sedimentological techniques are used to distinguish deposits of varied origin. Measures of particle size and shape help to
distinguish between bed material, point bar and levee deposits. Petrographic studies give indications of the geographical origin of a
material by taking account of the percentage of fragments of
different rock types of a given size, the variation of this percentage
as a function of size, and the nature of the heavy minerals in the
material. The stratigraphy of a deposit can provide a relative
chronology, while the presence of fossils or datable material may
enable the relative chronology to be fixed in time by accurate dating
of key horizons.

Lakes and deltas may be relatively short-lived features of the landscapes, in that lakes are infilled or drained and deltas are eventually
fully colonised by vegetation and become part of the general alluvial land surface or are drowned and eroded by a rise of lake level.
Nevertheless the behaviour and character of the lake while it exists,
the deposits on the lake floor and in the delta and the morphology
of the lake basin and its shoreline all provide some of the most useful information required by the geomorphologist to unravel the
history of the landscape.

IX

TIME AND THE
MORPHOLOGICAL UNITY
OF THE DRAINAGE BASIN

In the discussion of the effects of water movement over the surface of the land, the drainage basin has been considered the functional geomorphic unit. As Gregory and Walling (1971) point out, the topographic form and pattern of a drainage basin are standard characteristics inherited from the past. Consequently, the present processes changing the landscape may be seen as, in part, a function of landforms created under different sets of processes. The form of the drainage basin, produced over a long period of time, can influence the rate and the nature of the processes that operate within the basin system at the present time.

The drainage basin has to be seen as the product of several scales of landform evolution, ranging from small scale changes produced by individual storm events (order 7, Table 1) to the macro-scale features that result from tectonic disturbances in the geological past (order 4, Table 1). Few places in the world have such tectonic and climatic stability that drainage basin form can be said to be entirely due to the processes presently operating in the basin. All but the smallest drainage basins are therefore complex entities with a mosaic of relics from previous phases of landform evolution. Soils have in part begun to evolve under one climate to become dissected and eroded under another. Long established drainage lines are interrupted by earthquakes and constant drainage re-adjustment occurs. Removal of superficial sediments leads to the exposure in the channel bed of underlying strata which may be inclined at completely different angles to the overlying rocks. In the end, the river channel pattern within the drainage basin may become discordant with the geological structure of the country. Rivers that cut across the grain of the rocks of a region, not following structural trends, are discordant, the product of more than one phase of landform evolution.

Thus, if we hope to explain humid landforms, we need not only to understand the various processes involved in humid landform genesis, but to be able to unravel, for any given area, the roles played by different sets of past processes in the evolution of the drainage basin. This chapter examines the levels and time scales at which such understanding is required.

Previous chapters have discussed slope forms and channel patterns, making a distinction between the diffuse and unconcentrated actions of water and the work of water as open channel flow. The landscape does not provide such a clear-cut contrast. Hills and valleys, not slopes and streams, are the features perceived by the human eye. Thus, to discuss humid landforms realistically, valley forms must be examined. Louis (1964) suggested that valley forms may be associated with particular types of morphogenesis, but examples of most of his categories (Fig. 49) may be found in tropical rain forest areas.

28 *The feral relief and* Kerbtal *valleys of the southern flank of the Saruwaged Range, Papua New Guinea. (Crown Copyright reserved. By courtesy of the Director of National Mapping, Commonwealth of Australia.)*

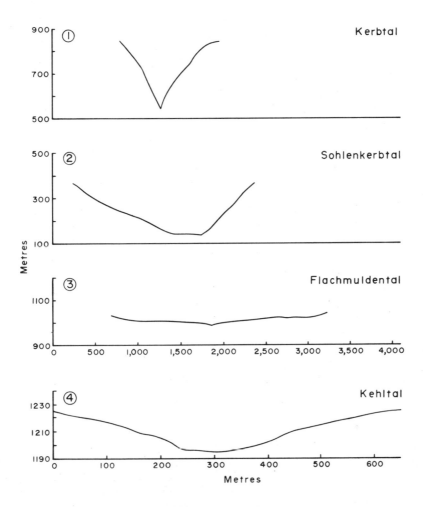

49 *Valley forms according to the classification derived by Louis (1968).
Valley A is Neck Creek in the feral relief of the New England gorge
country near Enmore, N.S.W., the term* Kerbtal *being used to describe
a V-shaped valley; B is Back Creek near Glenreagh in the hills between
Dorrigo and Grafton, N.S.W., the word* Sohlenkerbtal *describing a
steep-sided valley with a flat floor; the* Flachmuldental *of valley C, is
the shallow, saucer-shaped valley of Saumarez Creek on the New England
Tableland near Armidale N.S.W.; and Valley D is the* Kehltal *of the
Lohira River in Tanzania described by Louis (1964) as a gutter-shaped
valley, in that it has a generally rounded form like a gutter for carrying
water off a building.*

The feral relief (see Chapter VI and Plate 7) of mountainous, dissected tropical environments, such as New Guinea, central America, Main Range of West Malaysia and the slopes of the Bartle Frere massif of north-east Queensland, is characterised by steep, knife-edge ridges and valley slopes which meet at the river channel producing *Kerbtal* valley forms (Plates 28, 29). Further down valley these streams eventually develop a flat floor and a *Sohlenkerbtal* form in which the slopes descend steeply to the main valley floor, which is broad and swampy in places. In granite country, such as the Malaysian and Queensland localities mentioned above, and on arenaceous rocks, for example in central Pahang, flat valley floors are common, but may give way downstream to narrower and steeper valley forms, depending on lithological influences.

The basalts of the Atherton Tableland of north Queensland have *Sohlenkerbtal* forms which extend almost to the heads of first order streams. Valley sides have rounded summits but become straight and meet the valley floor at an abrupt angle (Plate 30). In terms of the 9-unit land surface model these basalt valleys lack units 4 and 6,

29   *Rapid incision of a* Kerbtal *on the north side of the Torricelli Range, Papua New Guinea. Lateral river erosion has caused a small mass movement (CSIRO photo)*

30 Sohlenkerbtal *on the Atherton Tableland, north Queensland*

there being a direct transition from unit 3, the creep zone of the slope, to a transport slope (unit 5) and thence to the valley floor. The absence of the colluvial footslope (unit 6) is particularly noteworthy.

Large valleys of this type have developed along major rivers in Hawaii where chemical weathering is intense, but because the voicanic rocks are extremely permeable only the deeper valleys maintain perennial springs. Major valleys grow at the expense of others as they tap larger supplies of groundwater. As weathering is most intense at, or just below, the water table, decomposition of material is most rapid near the floors of the deeper valleys, which are thus more easily deepened further when streams erode the weathered debris.

Physical breakdown and removal of debris from the fine grained basalts on the dry upper slopes is slow. Concomitantly, intensive chemical disintegration at the water table produces and maintains steep walled valleys. Owing either to sapping by weathering of the basal slopes or to lateral erosion by streams, valley floors tend to be flat. Where small accumulations of debris form small unit 6 footslopes, valleys tend to be U-shaped in cross-section. Units 2 and 3 are also poorly developed.

Small *Sohlenkerbtal* valleys in granite terrain are maintained by a combination of throughflow and flood erosion effects. Throughflow emerging at the foot of the slopes washes fine colluvial material out of the regolith and redistributes it over the valley floor. Frequent flooding of the valley floor spreads the sediment uniformly, carrying finer particles downstream. As the flow of water from upstream probably exceeds lateral inflows of water and sediment, the flat floor is maintained.

Higher order valleys in types of terrain just described and in less rugged environments tend to have less steep valley-sides and thus more of a *Kehltal* form. On older land surfaces, such as the Nuwara Eliya area of the hill country of Sri Lanka, broad flat *Flachmuldental* features occur.

In temperate humid forest environments all these types of valley occur. On the New England plateau of New South Wales, both *Flachmuldental* and *Kehltal* valleys are frequent, while to the east of the plateau, *Kerbtal* and *Sohlenkerbtal* types dominate the landscapes of coastal streams.

Valley forms are a function of slope and channel processes operating on varied substrates. In Hawaii, where base flow is high and stream activity great, flat floored valleys result from the decomposition and redistribution of stream sediments. Valleys where removal of material from the floor is not as effective as supply from the slopes acquire a more V-shaped cross-section. In high order humid tropical streams, lateral migration of meanders across the valley floor may remove detritus, most meandering rain forest streams having a rapidly shifting meander belt occupying the whole of the valley floor. However, dependent on the local lithological and structural conditions, areas where valley form is modified by abundant lateral supply of material from slopes, especially through mass movement and landslips, will be found. In the examination of valley and drainage basin morphology, the concept of the balance between slope and channel processes has to be borne in mind.

Frequency and magnitude of adjustments to drainage basin morphology

As Pain (1969) and Pain and Hosking (1970) have indicated in the Orere catchment of north-east New Zealand, which experiences high intensity, late summer, tropical cyclonic rainstorms, high magnitude storm events may control the evolution of drainage basin morphology. In the Hunua Ranges of the Orere catchment, extreme climatic events trigger mass movements which set off a wave of sediment transport through the channel system. Mass movement influences drainage basin morphometry by adding new first order segments to the channel network. The rare but high magnitude storms provide discharges which carry and eventually deposit large volumes of debris, thus creating floodplain deposits, which are not shifted until a storm of similar magnitude again occurs. A single high intensity storm may thus so disrupt the channel system that its effects on drainage basin morphology may be felt for many years after the initial event.

The lag between hillslope and channel processes

Knox (1972) has suggested that abrupt changes in hydrologic regimes related to biogeomorphic events in the drainage basin control the evolution of the channel and the floodplain. The evolutionary process can be reasonably described by a square wave lag function. In catchments less that 30 km^2 in area of south-western Wisconsin, where mixed hardwood forest gives way south-westwards to prairie vegetation, changes in climate produce marked changes in basin morphology and channel behaviour. Man's disruption of the hydrological cycle is but the most recent phase of a series of adjustments. In these fifth and sixth order valleys three distinct sedimentary sequences are found. At the base is coarse textured material, probably bed load sediment of a prior channel active towards the end of a mid-Holocene drought about 6000 years B.P. (before the present). Above this coarse layer is silty clay from a floodplain deposit created during the more humid conditions of the late Holocene. Distinctly laminated silt loam sediments laid down above the silty clay result from the increased surface runoff caused by modern land use practices. Climatic and land use changes of this type, expressed by changes in the vegetation cover, thus create disunity between channel and hillslope processes. Many temperate zone channels, particularly those of south-eastern Australia, such as the Tea-Tree rivulet in Tasmania (Goede, 1966, 1972), are today cutting into valley floor deposits derived from past phases of hillslope evolution (Plate 31). The material being excavated by rivers in these areas today is largely the product of a past phase of weathering and hill-slope erosion. The valley floor has acted as a temporary storehouse of debris, and erosion of the valley floor under these circumstances is not directly related to contemporary hillslope evolution.

31　*Strike-a-Light Creek on the Southern Tablelands of New South Wales, a stream actively cutting into valley floor deposits derived from earlier phases of hillslope development on the Tinderry Range*

THE NATURE OF TIME IN GEOMORPHOLOGY

Extreme events and lags between slope and channel processes raise the issue, already apparent from discussions of terraces and other depositional landforms, of the role of the past in the present landforms of humid regions. Evidence of previous processes from soil and weathering profiles and from floodplain, deltaic and lake sediments has already been mentioned. Such evidence may be extended by dating techniques such as carbon-14 (Thom, 1973), dendrochronology (Ralph and Michael, 1969; Armstrong, 1972; Flenley, 1974), varve counting (de Geer, 1934), tephrochronology (Pullar, 1967) and the analysis of prehistoric artifacts (Mulvaney, 1969; Vita-Finzi, 1974). Additional evidence of past conditions is provided by palynology (Colinvaux, 1973; Flenley, 1974) and palaeotemperature measurements (Broecker, 1965; Emiliani, 1971; Thom, 1973). Careful cross correlation and checking of dating and evidence has enabled much landform history to be written, but many general theories still rely on extrapolation of evidence from a few well-documental sample sites. Even the chronologies and patterns of environmental change derived from these sample sites are only as accurate as the dating and sampling techniques themselves.

The evidence of past changes does permit some conclusions about the order of magnitude of change in geomorphology. In landform evolution, time spans may be measured in terms of historical time or geological time. Historical time may range from a few minutes to several thousand years, from 10^5 to 10^3 years, a mere fleeting moment in geological time which is concerned with millions of years. In terms of the work done in transforming the landscape, identical time spans may be of unequal value, long periods of quiescence and relatively little landscape change being interspersed with sharp surges of activity during which considerable work is done (Tricart, 1965a).

Mass movements illustrate the contrasting values of time in geomorphology. A landslide involves three successive phases:

a period of preparation, which may last thousands of years, during which processes of weathering and erosion gradually create a potential disequilibrium. A stream undercutting a cliff may steadily increase the instability of the slope above, or underground water movement may so enlarge a karst cave that the roof of the cave begins to become unstable; *a brutal disruption*, which may last only a few minutes in the case of a landslide, or three weeks during a huge Alpine mudflow. During this stage vast landscape changes occur, but the immediate causes of the landslide appear small in relation to the total quantity of geomorphic work done. In reality, the event producing the landslide is but the fuse which sets off the results of the long periods of preparation. Unusually heavy rain infiltrating through cracks in a dry soil may be sufficient to lubricate an unstable mass along its slip plane and thus to set in motion vast quantities of material. The heavy rain is merely the catalyst permitting the reaction to take place; *a second period of slow modification*, almost imperceptible, during which the traces of the catastrophic event are steadily removed. Gradually the mass settles and is eroded by surface runoff, throughflow and solution.

The role of the brutal disruption phase in fluvial landscapes is well illustrated by the Orere catchment example described earlier in this chapter and by the effects of cyclones and hurricanes on river channel patterns and floodplains mentioned in Chapters VII and VIII. However, in many humid environments where the supply of material from slopes is not so abundant and not so affected by rapid mass movement, the rare or infrequent events of great magnitude may not be so significant. Particularly in relatively stable areas where heavy rainfalls are frequent, as in the humid tropics, analyses of the transport of sediment by various media indicate that a large portion of the work of erosion is performed by events of moderate magnitude which recur relatively frequently rather than by rare events of unusual magnitude (Wolman and Miller, 1960). Nevertheless, it must be recognised that *geomorphological time is heterogeneous, of unequal progression* (Tricart, 1965a). The changing significance of magnitude and frequency with type of sediment transport shown in Table 16 illustrates the point made in Chapter VI that the critical force required to initiate sediment motion varies with grain size.

TABLE 16 **Degree of mobility of material in rivers**

Grain size	Transport process	Frequency	Distance travelled
Ions	Solution	Permanent	Total
Colloids, etc.	Suspension	Seasonal	In part total
Fine sands	Mechanical suspension	Floods several times a year	Several km each time
Coarse sands	Bed load	Floods	Several 100 m each time
Pebbles	Bed load	Major floods	Several m or 10 m each time

In the study of changes in geomorphic processes with time, it is necessary to take account of:

the variations of the intensity of a phenomenon, which may or may not have a seasonal rhythm, such as the floods of seasonally wet tropical climatic regions;

the threshold values for the movement of given types of material;

the significant factors in the operation of processes, which may include particular combinations of independent variables, such as wind, runoff and gravitational stress.

In addition to these variations in historical time, longer term climatic variations must be considered. The two and a half million years of the Pleistocene have been marked by world-wide climatic changes associated with the expansion and contraction of polar ice caps. During the Tertiary warmer conditions than the present prevailed in temperate latitudes, giving rise to landforms quite different from those now evolving in such latitudes.

PLEISTOCENE LEGACIES IN DRAINAGE BASIN MORPHOLOGY

The climatic changes of the Quaternary have influenced most of the world's drainage basins. Glaciation above a snowline high in tropical mountains and declining to sea level in polar regions was accompanied by a world-wide shift of the climatic belts. During the Pleistocene after lesser oscillations, there appear to have been at least four major glacial episodes which can be traced from deposits in central North America, with indications of a fifth in Alpine areas. Within each of these episodes and during the interglacial periods between were lesser climatic fluctuations, Garner (1974) suggesting that probably some 87 to 111 second order climatic changes occurred during the Pleistocene Epoch. Such a series of changes implies an average of one climatic shift in every 25,000 years.

The glacial periods saw the lowering of the level of the sea and an increase in the extent of the land areas. Increased continental size would tend to create more marked continentality of climate in the interior of the land masses. The cooling of the earth's surface by about 5°C in the equatorial regions, but by much greater amounts in parts of the northern hemisphere (Markov, 1969), and the associated lowering of surface sea temperatures, would have led to a reduction in rates of evaporation over the oceans and a consequent decrease in rainfall over the continents (Galloway, 1965). Reduction of precipitation would have produced changes in the vegetation cover, often a change from a rain forest to a less dense plant cover, which, if rainfall was concentrated into seasons of heavy storms, would not have protected the soil against erosive raindrop impact, thus allowing coarse material to enter river channels and be transported along channels increased in length by the fall of sea level. In areas where now warm humid climates decompose feldspars in situ, coarse sand grains were eroded and redeposited. In Africa, the Nile, Congo, Niger and Senegal all deposited coarse arkosic material in the middle reaches of the present-day river. At times, during glacial periods, these rivers hardly carried enough water to flow through to the oceans, the Senegal having been cut off from the ocean by a dune barrier at one stage (Michel and Assemien, 1970).

Impact of Quaternary changes in the tropics

While broad latitudinal shifts of the major climatic zones appear to have occurred in Africa (Büdel, 1969), the north-south trend of the Andes greatly influenced Pleistocene climatic changes in South America. Much of the present lowland rain forest area east of the Andes became arid during glacial periods, though humid conditions and forest vegetation persisted at certain moderate elevations on isolated uplands and mountains such as those of Guyana and Imeri (Garner, 1974). Tricart and Cailleux (1965) have also suggested that coastal areas, islands and peninsulas, including parts of Indonesia and the south-west of Sri Lanka, would have remained rain forest refuges, while the continental interiors became more arid.

Ice caps formed not only in high latitudes, but also on high tropical mountains. In the tropical Americas the snowline was 1500 m lower at 30°S and about 700 m lower at 12°S during cold periods, with the lowering being greater on the Pacific than on the Atlantic side of the Andes (Hastenrath, 1971). Associated with this lowering of the snowline were higher rainfalls in montane regions. So while the lowland areas of the tropical interiors were more arid than at present during glacial periods, parts of the montane tropics may have been wetter. Thus in northern Chile, Peru, Colombia and north-west Venezuela, pluvial periods coincided with glacials; in Guyana, Surinam, the middle Amazon and the Argentine Pampa, the climates of glacial periods were drier than at present (Moushinho de Meis, 1971; Tricart, 1969; Zonneveld, 1968).

The Indian sub-continent shows a similar coincidence of dry periods in equatorial areas with the Himalayan glaciations (Joshi, 1972; Brunner, 1970). Probably there was less rain in the Deccan during glacial periods than at present, while the interglacials experienced wetter conditions than those now prevailing. During the pluvials deciduous tropical forest extended into areas which are now savanna, while the arid periods saw reduction of the vegetation cover with related changes in runoff and sediment yield.

These Quaternary climatic fluctuations raise the issue of the duration of one set of landforming processes, or even of the vegetation and soil components of the denudation system. Charcoal layers under supposedly undisturbed Nigerian forests, archaeological remains under Central American and South-east Asian forests, and the recorded rates of forest recolonisation of Indonesian volcanoes all indicate how rapidly tropical evergreen forests regenerate after disturbance. Nevertheless, until stratigraphic and palynological studies are avialable from a wide variety of equatorial areas, it will remain impossible to clarify the nature of Quaternary climates in the humid tropics.

Sea level change and humid tropical river systems

The impact of such climatic changes on landforms stems from the alternation of process types. In the middle Amazon, the lower sea level of glacial periods caused rivers to cut down towards a lower base level and to develop longer, steeper major channels. Increased aridity and poorer vegetation cover, with less frequent runoff, caused mass movements of unprotected soil to fill low order stream channels, drainage density probably decreasing as the supply of sediment from the slopes exceeded the capacity of stream discharge to carry the detritus along the river channels. However, in the postglacial period, streams began to aggrade their channels as sea level rose while climatic conditions remained dry. The large spreads of alluvium of tropical countries, such as Malaysia and Borneo and Sumatra, relate to deposition during fluctuating sea levels. The older alluvium of Johor and Singapore is mid-Quaternary material deposited at a high interglacial sea level. It was subsequently dissected during a glacial, and the present meandering coastal streams occupy zones of postglacial aggradation. The middle Amazon today presents a contrast between the streams rising outside the lowlands, which carry large quantities of sediment from the Andes (Gibbs, 1967) and are continuing to build up their beds, and the streams of the lowland itself which are gradually being drowned (Moushinho de Meis, 1971). Similar characteristics are to be found on the extensive coastal plains of South-east Asia.

Quaternary sea level changes (Bird, 1968) and related adjustments of the lower courses of river channels mean that interpretations of landform history based on forms alone (i.e. solely on morphological evidence as deduced from field survey or from maps or aerial photographs) may be fraught with difficulty. Geologists in Malaya in the first quarter of the twentieth century studied the stream network observable on the map and suggested that the Pahang River, which flows from north to south in its upper and middle reaches before making a right angle bend to the east at Temerloh, once continued more or less directly southward to the present Muar River, which reaches the sea on the west coast of Johor through a series of large meanders. Ho (1962) suggested that the Pahang's lower course evolved during the fall of sea level from the height at which the older alluvium was deposited and that the high sediment load carried from the northern mountainous tributaries blocked off virtually all the southern tributaries, creating the Tasek Bera, much in the way that the present-day river levee cuts Tasek Chini off from the Pahang. The large meanders of the Muar River are produced by the evolution of the channel in mangrove swamps during fluctuating sea level conditions, with possibly tidal action of the type described by Geyl (1968).

As the Torres Strait was a landbridge at the time of glacial low sea level, conditions in southern Papua may well have been more arid, certainly more continental and thus probably more like northern Australia today.

Changes in climate produced differences in slope processes, creating the alternating slope deposits found in southern Papua (Mabbutt and Scott, 1966). The same changes could partially explain some of the shifts between sclerophyll and rain forest vegetation in north Queensland revealed by palynological studies (Kershaw, 1975).

With such powerful changes occurring in the tropics during the Pleistocene, it cannot be supposed that all, or even most, of the landforms under tropical rain forests were produced by the processes operating at present. For many areas there is no evidence to suggest that present-day processes have been able to do more in the last few thousand years than to decompose some of the Pleistocene debris and to recolonise erosional and depositional features produced under less warm and humid conditions than those prevailing in the humid tropics today. Much more evidence is needed from both terrestrial and marine deposits before we can fully evaluate whether even the humid landforms of the tropical rain forest can be said to be the 'normal' landscape created by the action of running water.

THE QUATERNARY LEGACY IN TEMPERATE HUMID LANDS

If the Quaternary changes in the humid tropics outlined above were drastic, those in the now humid temperate lands were even more so. In the northern hemisphere ice sheets extended into southern Britain, middle North America and well over most of northern Europe. Beyond the ice sheets spread stretches of fluvio-glacial deposits and beyond them large areas were frost-dominated wastelands which left a large quantity of periglacial debris on the hillslopes and valley floors of middle latitudes, as well as in the higher areas of lands in North Africa, and such southern hemisphere parallels as south-eastern Australia.

Vegetation reacted profoundly to the climatic changes in the northern hemisphere, but there was no general orderly retreat in latitudinal zones. The postglacial succession in the New England region of the U.S.A. shows a cold (boreal) start with conifers predominating, with, in some cases, short periods of treeless tundra conditions immediately after the retreat of the glaciers, then the development of the broad-leaved forest which remained until European settlers arrived in the seventeenth century and started to cut the forest. The coniferous trees seem to be as much a part of the development of the ecological succession of the vegetation as the product of the previous existence in the area of a boreal forest latitudinal belt (Colinvaux, 1973). In the early part of the postglacial of the Dismal Swamp of Georgia there are also clear signs in the pollen record of northern conifers like spruce (*Picea*) being present, but these occur with a different set of bedfellows. Palynological evidence suggests that glacial periods produced a resorting of the species of the American forests, so that combinations quite unlike those of the present day existed.

Role of relict debris from past climates in present-day landforms

The significance of changes in vegetation type during the Quaternary for the study of humid landforms lies in the alternation of forest structure and the changes in forest hydrology. Different ecological situations imply changes in the circulation of chemical elements and in weathering processes. Such changes in process mean that today the material being carried by streams is often the historical hangover of material prepared by weathering and shifted by mass movement at various stages in the past. Kirkby (1967) discovered this in Scotland; Douglas (1967b) noted this in the Southern Highlands of New South Wales. Valley floors are filled with debris from a past climatic phase. Debris is supplied to the streams largely by bank erosion, while the colluvial deposits on the slopes are often the legacy of a past phase of operation of the denudation system.

Many areas that were covered by ice, or lay at the margins of the ice sheets, are covered with glacial drift, sometimes called till or boulder clay. Such materials blanket the local details of the preglacial topography so that the landforms seen today are of composite origin with elements of different ages. Thus, while it is possible to study the mechanics of rivers in temperate environments, and to obtain fundamental data on the forces involved in river mechanics, it is not possible to develop an adequate theory of landform evolution through geological time from the study of present-day humid temperate landforms. This statement deliberately disagrees with the comment by Carson and Kirkby (1972) that it is difficult to understand why any appreciable modification of the *form* of slopes in profile should have taken place as a result of Pleistocene periglacial processes. These authors argue strongly that many hillslopes stand at angles which are apparently entirely explicable in terms of the type of material mantling the slope and that it is quite possible that the basic geometry of the slopes was determined under humid temperate, or other, conditions before Pleistocene glaciation.

Despite these assumptions, largely based on British evidence, soil stratigraphic evidence, such as that from Nowra described in Chapter III, clearly indicates that different soil mantles have accumulated at various times in the Pleistocene. As soil development takes time, a soil mantle represents a stable surface, a period of relative still stand in landscape evolution. The soil mantle is thus one phase of an erosion-deposition sequence with the period of active change eroding material from one part of the slope and redepositing it in another.

Studies of soil mantles at a variety of localities in south-eastern Australia (Butler, 1963, 1967) indicate that despite different causes of the initiation of erosion and possible lack of contemporaneity between localities, general hillside erosion and hillfoot deposition has been the common event, except, of course, on the plains, where riverine deposition and accretion of fine aeolian clay known as parna dominate. Definite periodicity of landscape evolution has been recognised and the hillslopes retain mantles of differing Quaternary age.

Quaternary adjustments of river systems

In addition to changes in river systems due to sea level fluctuations, climatic changes produced changes in regime which affected channel patterns and channel sediments. The heads of channels may have been infilled during periods of active slope wash but infrequent channel flow, sheet wash replacing concentrated flow. Such changes in drainage density would have important consequences for the evolution of channel patterns. Phases of entrenchment and alluviation during Quaternary climatic changes have been discussed in Chapter VIII.

Many meandering valleys contain floodplains on which present-day rivers describe free meanders of much smaller amplitude than the enclosed valley bends. Streams in such situations are commonly called *underfit*, the implication being that the windings of the valleys are authentic meanders — frequently, ingrown meanders — of former large streams, that there is a relationship between stream size and size of meanders, and that some cause has operated to reduce stream size and meander size significantly below their former values (Dury, 1970b). As underfit streams occur in many parts of the world, including all the major climatic regions of the coterminous U.S.A., Europe as far east as the Ukraine and in the eastern third and Northern Territory of Australia, a general theory to account for their existence has to be broadly applicable and most probably is related to climatic change.

Dury's detailed investigations (1965, 1970b) of underfit streams indicate that with the warming up since about 9000 B.P. revealed by palaeotemperature measurements (Fig. 50) and other fossil evidence, many temperate areas became drier. The reduction of dominant discharge saw floodplain infilling and the development of channels with reduced meander wavelength. Controversy surrounds Dury's conclusion that the channel forming discharges required to explain the former channel patterns of streams which are now underfit must have been provided by increases in mean annual precipitation of 50 to 100 per cent. His studies imply an increase in storminess, not only in mid-latitudes but also beyond, during the last episode of deglaciation. The state of the global circulation during maximum glaciation still remains uncertain. In Australia, stronger wind circulations and greater volumes of snow during glacial periods (Derbyshire, 1971) are considered to have been a possible source of water for increased stream flows during seasonal snow melt events. However, insufficient is known of global perturbations of the general circulation to ascertain whether similar

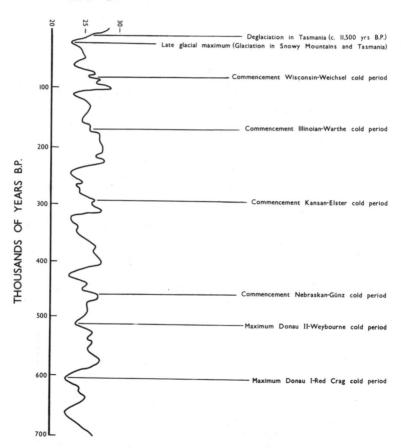

50 *Generalised palaeotemperature curve based on all available isotopic records (after Emiliani and Shackleton, 1973). Approximate dates of cold periods are indicated.*

causes apply elsewhere. Probably the phenomenon of reduction in dominant discharge occurred several times during the Quaternary, and the present underfit streams are but the latest in a series of adjustments of streams to river regime changes.

The depositional channels of the Murray-Darling basin of southeastern Australia provide a natural laboratory for the testing of the principles of hydraulic geometry. The Riverine Plain of southeastern Australia has a reasonably well understood geological history for the last 50,000 years. Apart from secondary aerolian features, the entire plain, which covers 65,000 km^2 of New South Wales alone, consists of water transported alluvial and lacustrine sediments by inland flowing river systems. Depositional streams fanned out from three gaps in the ranges of New South Wales and Victoria, at Hillston, Narrandera and Corowa. In the eastern half of the plain, nearer the apices of these fans, stream traces are closely spaced, streambed and levee deposits predominating over the floodplain components of the landscape. On the western part of the plain, former channels are more widely spaced, with less developed levees, so that floodplain deposits predominate. The lakes in the far west of New South Wales were probably the product of impeded flows in distributary streams, the present lakes being superimposed on former and often much larger lake remains.

The younger set of the earlier river channels are termed 'ancestral river channels'; the older set 'prior stream channels'. In pattern, shape and sediment characteristics, the ancestral river channels are similar to the modern river channel, but are much larger. The prior stream channels, on the other hand, are relatively straight, wide and shallow, being filled with cross-bedded sands. The differences between these channels result from contrasts in discharge regimes of these rivers during changing Quaternary climates.

Although the modern Murrumbidgee River transports little sediment, the prior streams carried large quantities of sand across the plain, and were characterised by higher flood peaks, but lower mean annual discharge than at present (Table 17) (Schumm, 1968). The ancestral streams had greater annual runoff and higher flood flows than the present river channel, which suggests that the ancestral channels were formed under more humid conditions than the present, while the prior streams existed in conditions more arid than the present.

TABLE 17 **Morphology of riverine plains channels**
(after Schumm, 1968)

Location	Murrumbidgee River	Paleochannel 1	Paleochannel 2
Channel width, in metres	67	140	183
Channel depth, in metres	6.4	10.7	2.7
Width-depth ratio, F	10	13	67
Sinuosity, S	2.0	1.7	1.1
Gradient, S, in metres per km	0.13	0.15	0.38
Meander wave-length in metres	853	2134	5490
Median grain size in millimeters	0.57	—	0.55
Channel silt-clay, M, in per cent	25	16	1.6
Bedload Qs, in per cent	2.2	3.4	34
Bankfull discharge, in $m^3 sec^{-1}$	594	1443[a]	651[a]
Sand discharge at bankfull in tonnes per day [b]	2000	19 000	49 000

[a] Calculated by use of Manning equation and channel area.

[b] Calculated by Colby's technique.

The complex of depositional landforms produced by the prior streams is possibly too old to date by radiocarbon methods, i.e. older than at least 36,000 years B.P. (Pels, 1971). However, after the complex was formed, periods of stream incision and valley fill have occurred on the Riverine Plain. They correspond to periods of erosion and relative stability on the Southern Tablelands. Costin (1971) produces evidence for a fairly general cooling off of the Southern Tablelands and Highlands beginning about 32,000 years ago. It may have remained cold, with mean temperatures 8·3° to 11°C lower than the present, until about 20,000 years ago. However, by 15,000 years B.P. the ice and snow cover had disappeared sufficently to permit vegetation similar to that of today, with conditions improving further after 10,000 years B.P. after which conditions of climate, soils and vegetation appear to have been generally similar to those at present, save for a colder period 3000 to 2000 years ago.

This record of events in the headwaters of the Riverine Plain drainage basins has to be compared with the record of depositional and channel fill phases established by Pels (1971). Degradation by streams took place approximately in the periods 30,000 to 26,000 B.P., 13,400 and 10,000 B.P. and 4000 B.P. to the present. These would coincide with the onset and end of Costin's main cold period and the less rigorous cool period beginning 4000 years ago. However, the analysis of the evidence from upland valleys and the depositional plain is incomplete, and the full story of the evolution of the Murray-Darling basin as a morphological unit is yet to be written. Did the high discharges causing channel incision occur during the phases of changing climate when instability of climatic regime may have been greatest? Can the depositional sediments be related to ground surfaces and soil mantles on the hillslopes?

Difficulties in obtaining precise information on Quaternary events

While it is possible that the Quaternary processes only produce landscape modifications on the details of hillslopes, valleys and streams, and that the gross forms were inherited from the Tertiary, the timing, dimensions and nature of Quaternary changes in geomorphic processes are imperfectly known. Commenting on evidence for late Quaternary climates in Australia, Galloway (1971: 22) comments:

> It is painfully clear that we really know very little. From the evidence of glacial and periglacial features, sea bed fossils and possibly oxygen isotope studies, we know that lower temperatures existed in southern Australia and Tasmania at a time corresponding to the last great glaciation of the Northern Hemisphere. We do not know how much lower the temperature was nor if there were regional variations in the amount of lowering. Abandoned lake shore features show that the ratio of evaporation to precipitation has been much smaller than it is today but we do not know

if actual precipitation was more, about the same, or less. We have no definite evidence for climatic fluctuations within the last glaciation other than some scrappy observations in the very limited glaciated areas of the continent and one or two deep sea cores.

Faced with these problems, Derbyshire and Peterson (1971) point out that the establishment of a regional and interregional stratigraphy for the Australian Pleistocene is most likely to come from the former periglacial and extra-glacial areas, and, with the growth of oil exploration and oceanographic research in the seas around Australia, from the continental shelf, including its marginal plateaux, and the abyssal plains. It is clear, however, that judgments about the relationship of landforms in humid temperate areas to specific climatic and ecological conditions must be full of caution until local Quaternary histories are established.

PRE-PLEISTOCENE CHANGES IN DRAINAGE BASIN MORPHOLOGY

Both climatic and tectonic changes have affected landform evolution throughout geological time. Even the drifting of the continents and the influences of plate tectonics may still be reflected in the evolution of present river systems. Earthquakes may produce abrupt changes in river courses, while fault systems control the paths followed by rivers in many cases. In the Tertiary, climate was generally stable, and long periods of relative stability of landscape existed in tectonically inactive areas.

Tectonics and humid landforms

Over much of Australia, including the now humid east coast areas, weathering crusts were formed, but some tectonic activity occurred at intervals during the period. Faults were activated, and reactivated later. Basalts and volcanic lavas were ejected on to much of what is now the Eastern Highlands. Faulting and basalt outpouring affected the river courses considerably. Major tectonic uplift created the great waterfall and gorge systems from the Shoalhaven to the Barron River. Basalt flows filled up valleys, giving rise in places to the buried 'leads' of mineral-rich alluvium such as those of the ancestor of the Wild River near Herberton in north Queensland. Often new streams developed around the edges of basalt flows, or centrifugal first order streams radiated from a volcanic crater such as Mount Warning in northern new South

Wales (Fig. 43). In most cases basalt flows have produced only local modifications of drainage alignments. Headwaters of some streams have been truncated, their waters being diverted over former divides, as happened to the head of the Walsh River in the vicinity of Mareeba in north Queensland.

A wide variety of tectonic influences have affected river systems. On the one hand, there are major rivers, such as the Amazon and Mississippi, which have discharged into the same, but steadily subsiding, sedimentary basin at least since the beginning of the Tertiary era. On the other hand, systems such as that of the Niger, with great inland deltas, show a complex pattern of evolution affected by both tectonic and climatic change. Even though river systems that evolved under more humid conditions than the present may now only be fully operational following the heaviest storms, their morphological unity is apparent, even if climatic or tectonic changes have modified their courses. Even the massive uplift of the Himalayas was matched by the incision of the antecedent Brahmaputra which cuts through the mountains in a gorge several thousand metres deep.

Tertiary climates of present-day humid temperate areas

The evolution of the oceans through the Tertiary was accompanied by movement of the positions of the poles and equator (Ahmad, 1973). Some of the evidence of warmer Tertiary conditions may be due to these shifts. Traces of Tertiary tropical climates in Europe suggest that large segments of the landscape may have acquired their gross morphology under a warmer, possibly seasonally wetter climate than that of the present. Tricart and Cailleux (1965) have suggested that from the middle to late Tertiary much of France may have had arid conditions, as several late Tertiary saline deposits are to be found. To the south-east in southern Germany, Hungary and Romania, the existence of wetter tropical conditions during the Tertiary has been invoked (Pop, 1964; Büdel, 1957b) while in Galicia, granitic rocks were decomposed under humid subtropical conditions in the late Tertiary (Godard, 1972).

Even though curves of mean annual temperature at sea level in central Europe from the Eocene to the present have been published (Woldstedt, 1954), with an indication that temperatures were virtually constant at 22°C through the Eocene and Oligocene, considerable doubts as to the nature of the climate at that time have risen from detailed studies of the flora of the London Clay of southern Britain. Although the London Clay contains some species associated with tropical rain forest and was taken to indicate a humid tropical climate in Britain in the Eocene, it now appears (Daley, 1972) that the rain forests were an extension of tropical flora into temperate latitudes, with the tropical species occurring as gallery forests in moist areas, with extra-tropical species in the less favoured sites, a vegetation combination that does not occur today.

Evidence of climatic change in the Tertiary emphasises the poly-genetic character of humid temperate landforms. Landforms on crystalline rocks and sandstones in Britain and France have clear indications of deep weathering under warmer conditions in the Tertiary and removal of the weathering mantle and detailed sculpting of minor features in the Quaternary (Linton, 1955; Godard, 1966, 1972). Similar sequences of events in Australia have been invoked for the granite tors of the tablelands of New South Wales and Victoria. Certainly deep weathering has occurred from the granite belt of the Stanthorpe area of southern Queensland to the Snowy Mountains of New South Wales and the weathered material has many affinities with that now forming on Singapore Island, yet does such deep weathering depend more on climate or on the nature of the parent material? Ollier (1965) has argued that the deep weathering of granite can occur in a wide range of humid climates. Dejou and co-workers (Dejou, 1967a, b, c; Dejou, et al., 1968, 1969, 1970) have shown that gibbsite and kaolinite, long thought to be indicators of humid tropical weathering conditions, are forming at present in the humid temperate climate of the Massif Central of France. Even more compelling is the evidence reviewed by Paton and Williams (1972) indicating that laterite-like material can form in cool temperate to subarctic climates given suitable parent lithology and rapid leaching.

The presence in Australia of large areas of duricrust has been taken to indicate the stability of a largely worn-down landscape of low relative relief in the Miocene (Browne, 1969). This duricrust is thought to be largely lateritic, as Browne refers to siliceous pisolitic crusts as distinct from the duricrust proper. This loose terminology confuses the traditional view of Australian geologists of Miocene Australia as having a 'widespread and conspicuous' land surface of low relief, with a 'wonderfully level skyline' on which developed a layer of lateritic material. The Lachlan formation in the Cowra district is of Miocene age and contains species related to both tropical and temperate rain forest conditions, which suggests that some of Australia may have been more humid in the Miocene that at present. However, just as detailed studies in Europe have raised questions about the uniformity of Tertiary conditions, so more recent inquiries in Australia are suggesting that a uniform Miocene climate was most unlikely and that the so-called Miocene duricrust may comprise elements of widely different ages.

Much of the evidence for Tertiary landscapes comes from parts of old land surfaces which have remained relatively immune from later erosion. These elements are often only small fragments of the present-day landscape and extrapolations from them must be viewed cautiously. However, despite the controversy over what happened through geological time, enough is known to indicate that a large part of the land surface in present-day humid denudation systems may have had considerably different climatic conditions in the Tertiary. In particular, not only are the details of humid temperate landforms largely the product of Pleistocene events, but their gross forms may be the product of denudation under considerably warmer Tertiary morphogenetic systems. Further complications arise when the drifting of the continents and orogenetic events are considered. The landforms of New Guinea are virtually late Tertiary and Quaternary, those of Borneo reflect the evolution of geosynclines during the Tertiary, while those of Sri Lanka are much older, being the etching out of a fragment of the ancient Gondwanaland continent. Tectonic changes create new local climatic circumstances with further feedback on denudation systems. Broad generalisations about climatic change are only part of the story.

Continual adjustment and modification of river basins proceeds, often at varying rates within the same basin, producing the complex landscape patterns containing elements of different ages and origins that give the land surface its great variety and which pose geomorphologists countless questions. Every detail of form requires careful examination, using all available knowledge of associated depositional features, residual soils and the nature and effects of fluvial processes, to be able to attempt to explain the pattern of evolution of that part of the river system. A good theory of landform evolution must take account of tectonic events, climatic change, and scale.

X

THEORIES OF
LANDFORM EVOLUTION

DYNAMIC GEOMORPHOLOGY

In the previous chapter it was pointed out that geomorphological time is heterogeneous, of unequal progression. Events affecting the shape of the earth, be they floods giving rise to movement of large quantities of sediment in rivers, or climatic changes causing vegetation alteration with consequent modifications through changes in erosion rates, or tectonic forces, from earth tremors to the more drastic earthquakes, such as that in Alaska in 1964, occur at irregular intervals. Such evidence that processes producing landforms vary in magnitude and intensity must be part of any general theory of landform explanation. This book has discussed landforms in terms of their dynamics, describing them as constantly changing features, responding to changes in processes, examining the ways in which these processes operate and how they are inter-related. If this material is brought together with contemporary knowledge of plate tectonics, continental drift, and mountain building, the existence of two broad groups of geomorphic processes can be appreciated: those operating internally, tending to distort and uplift the land surface, and those operating externally, tending to smooth off, to erode the most rugged areas and to fill depressions. These external and internal forces have been termed exogenetic forces and endogenetic forces respectively and can, within certain limits, be envisaged as operating against one another.

In 1952, Strahler urged geomorphologists to turn to the physical and engineering sciences and mathematics for the vitality the discipline then lacked, pointing out that geomorphic processes are basically the various forms of shear, or failure, of materials which may be classified as fluid, plastic or elastic substances, responding to stresses that are commonly gravitational, but may also be molecular. Processes should be studied and landforms characterised quantitatively so that they can be compared statistically (Strahler, 1952b). Previous chapters indicate how far understanding of processes and the quantitative approach to geomorphology have advanced since 1952. However, Strahler also urged the development of the concept of open dynamic systems and steady states for all phases of geomorphic processes, and finally the deduction of general mathematical models to serve as quantitative natural laws. This search for applicable systems concepts and general models continues. Before Strahler some good models of landform evolution existed. Since 1952 other models, both mathematical and non-mathematical, have been put forward. This chapter will attempt a brief review of their relevance to the study of humid landforms.

Geomorphological theories have to cope with the interaction of exogenetic and endogenetic processes with materials through time (Tricart, 1965b); they are involved with transformations of mass and energy as functions of time. This analysis of changes in the three-dimensional forms through time has much in common with four-dimensional equilibrium theories of mathematical physics, though lacking their sophistication and precision (Culling, 1957).

The consequences of transformations of mass and energy through time in river systems were described by Gilbert (1880), who pointed out that a stream of given size, carrying a given load, will develop a graded profile and will tend to maintain that profile. Gilbert also showed that if the load is reduced or the volume of water flowing in the stream is progressively increased, downcutting ensues, and the stream develops a new profile in response to the new conditions. While this type of adjustment of the longitudinal profile is one example of an equilibrium trend and the relevance of the regime concept, it is also an example of progress through time towards a final form. The most far-reaching and influential general model of landscape evolution takes into account the succession of forms through time.

THE CYCLE OF EROSION

'All the varied forms of the lands are dependent upon — or, as the mathematician would say, are functions of — three variable quantities which may be called structure, process, and time' (Davis, 1899).

From a study of the landform history of the plains of Montana, Davis developed ideas on the sequence of events leading to the creation of plains of denudation, cut down to old base-levels of erosion (Chorley, Beckinsale and Dunn, 1973). Over two decades he enlarged these ideas into the concept of the cycle of erosion, a theoretical model for describing landforms, not only as morphological features, but as features with a distinct phase of evolution, some being 'young' while others are 'mature' or even 'old'. If conditions are stable for long enough, the landscape would eventually 'be worn down smooth and low to featureless plain' (Davis, 1899).

The excellent series of articles accompanied by block diagrams which set out the ideas and applications of the cycle of erosion have been discussed at length by Chorley, Beckinsale and Dunn (1973) and only a few key points need emphasis here. Davis saw the cycle of erosion as a deductive model:

> The scheme of the cycle is not meant to include any actual examples at all because it is by intention a scheme of the imagination and not a matter of observation; yet it should be accompanied, tested and corrected by a collection of actual examples that match just as many of its elements as possible.

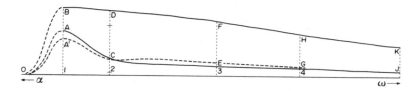

51   *The geographical cycle after Davis (1899). The symbols are explained in the text.*

Davis illustrated his fundamental postulates with a graphical model (Fig. 51) in which:

> 'The baseline $\alpha\omega$ represents the passage of time, while verticals above the base line measure altitude above sea-level. At the epoch 1 let a region of whatever structure and form be uplifted, B representing the average altitude of its higher parts and A that of its lower parts, AB thus measuring its average initial relief. The surface rocks are attacked by the weather. Rain falls on the weathered surface, and washes some of the loosened waste down the initial slopes to the trough-lines where two converging slopes meet; there the streams are formed, flowing in directions consequent upon the descent of the trough-lines. The machinery of the destructive processes is thus put in motion, and the destructive development of the region is begun. The larger rivers, whose channels initially had an altitude, A, quickly deepen their valleys, and at the epoch 2 have reduced their main channels to a moderate altitude, represented by C. The higher parts of the interstream uplands, acted on only by the weather

without the concentration of water in streams, waste away much more slowly, and at epoch 2 are reduced in height only to D. The relief of the surface has thus been increased from AB to CD. The main rivers then deepen their channels very slowly for the rest of their life, as shown by the curve CEGJ; and the wasting of the uplands, much dissected by branch streams, comes to be more rapid than the deepening of the main valleys, as shown by comparing the curves DFHK and CEGJ. The period 3–4 is the time of the most rapid consumption of the uplands, and thus stands in strong contrast with the period 1–2, when there was the most rapid deepening of the main valleys. In the earlier period, the relief was rapidly increasing in value, as steep-sided valleys were cut beneath the initial troughs. Through the period 2–3 the maximum value of relief is reached, and the variety of form is greatly increased by the headward growth of side valleys. During the period 3–4 relief is decreasing faster than at any other time, and the slope of the valley sides is becoming much gentler than before; but these changes advance much more slowly than those of the first period. From epoch 4 onward the remaining relief is gradually reduced to smaller and smaller measures, and the slopes become fainter and fainter, so that some time after the latest stage of the diagram the region is only a rolling lowland, whatever may have been its original height. So slowly do the later changes advance, that the reduction of the reduced JK to half of its value might well require as much time as all that which has already elapsed; and from the gentle slopes that would then remain, the further removal of waste must indeed be exceedingly slow. The frequency of torrential floods and of landslides in young and in mature mountains, in contrast to the quiescence of the sluggish streams and the slow movement of the soil on lowlands of denudation, suffices to show that rate of denudation is a matter of strictly geographical as well as of geological interest.

It follows from this brief analysis that a geographical cycle may be subdivided into parts of unequal duration, each one of which will be characterized by the degree and variety of relief, and by the rate of change, as well as by the amount of change that has been accomplished since the initiation of the cycle. There will be a brief youth of rapidly increasing relief, a maturity of strongest relief and greatest variety of form, a transition period of most rapidly yet slowly decreasing relief, and an indefinitely long old age of faint relief, on which further changes are exceedingly slow.

Despite the assumption of tectonic stability in this general model, Davis pointed out that:

the postulate of rapid uplift is largely a matter of convenience, in order to gain ready entrance to the consideration of sequential processes and of the successive stages of development, — a young mature and old — in terms of which it is afterwards so easy to describe typical examples of land forms.

He emphasised that the general model is a simplification put forward to assist comprehension, but that:

> instead of rapid uplift, gradual uplift may be postulated with equal fairness to the scheme, but with less satisfaction to the student who is then first learning it; for gradual uplift requires consideration of erosion during uplift.

Presenting an illustration of how slow uplift could be incorporated into the cycle of erosion, Davis (1905) described an almost steady-state landscape of broadly open valleys with gently sloping, evenly graded sides descending to the stream banks, leaving no room for a floodplain. However, he adds that uplift is usually much faster than downwearing and that: 'That original postulate of rapid uplift therefore requires only a moderate amount of modification to bring it into accord with most of the landforms that we have to consider'.

The writings of W. M. Davis have been quoted at length because they have had such a great influence on geomorphological thinking, particularly in Britain, and as he has been criticised recently, it is important to realise the nature of his own statements. However, it should now be clear that Davis set up a theoretical reasoned argument about the cycle of erosion which related forms to stages of landscape evolution.

Davis used the cyclic concept to interpret how erosion cycles in actual landscapes had been complicated and interrupted. As base levels changed, new cycles were initiated while the landscape preserved traces of older cycles. By recognising these traces of previous phases of landform evolution, it should be possible to reconstruct the geomorphological history of an area. This Davis did for many areas, but his finest work was probably the study of Appalachian erosional history, which he illustrated with a series of block diagrams showing the successive stages of evolution of the Appalachian landforms, from the peneplain to the adjustment of rivers to the structures of the rocks that are exposed as the peneplain is dissected. Such diagrams portray stages in the denudation chronology of the area concerned. Many workers followed Davis's example and used both the slope of the drainage network and the occurrence of level upland surfaces cutting material of different lithology as evidence of possible former peneplains to establish the denudation chronology of their own field areas.

Misuse of the Davisian model

In Australia, Taylor (1958) used the Davisian method to explain the scenery of New South Wales, describing the gorges of the Blue Mountains as 'some of the finest juvenile gorges in the world', the country around Parramatta as 'a late mature landscape'; and Cook's River, just north-west of Botany Bay, as an 'old stream'. Taylor's statement that 'the same river can exhibit different ages in different parts of its course, depending on contemporary varia- tions in the amount of elevation or depression of the land' probably illustrates an extreme case of misuse of the anthropomorphic stage names of the geographical cycle. While a gorge on the headwaters or middle course of a river may be geologically young, so also is the floodplain across which that same river meanders before entering its delta. The comments on the impact of changing Quaternary sea level in the previous chapter have shown how recent the aggra- dation of most floodplains must be. A river basin may contain landform elements of varied age, but it is quite wrong to assume that the age of a landform may be deduced from its appearance or morphology.

Unfortunately, Taylor's example has been followed by the geo- logists of New South Wales right up to the present day. Browne (1969) comments that:

> as the eastern rivers approach the coast their juvenile gorges in general pass into ever-widening valleys bordered by alluvial terraces or into broad, alluviated coastal plain resulting from Quaternary eustatic movements. In the west, owing to the generally lower topographical relief, valley-development has been more rapid, and in the western slopes the rivers have already attained a condition of advanced maturity and even senility.

This type of use of Davisian ideas does not advance the under- standing of how the landscape came into being. Criticism by Hettner (1928, 1972) of this type of use of stage names in geomorphology was countered by Davis (1923) that it is *stage*, not age, that is discussed in the geographical cycle, and, 'although stage of develop- ment manifestly depends on the factor of time, it is practically determined not by measuring the geological periods during which a given feature has been undergoing erosion but *simply by its visible forms'*. In the case of the streams affected by the uplift of the Eastern Highlands, which was far from the theoretical single tectonic movement of the ideal geographical cycle, the eastward and westward flowing rivers are of the same age. Admittedly they contain elements of pre-uplift river courses (Warner, 1971), but all their channels have adapted to changes in the base-level, to tectonic events, to the varied resistance of the rocks over which they flow, to Quaternary changes in the hydrological cycle, vegetation cover and pedogenesis. As a result these rivers are today showing varying degrees of adjustment to the present climatic and ecological conditions. The streams of the western slopes have abandoned the prior and ancestral channels which developed earlier in the Quaternary, east coast rivers have created new terraces below those which grade steeply towards the lower sea level of the past. Even if stage names apply to 'visible form' only, it is somewhat misleading to describe streams equally actively evolving at the present time as 'juvenile' on the one hand and 'late mature' on the other.

DYNAMIC THEORIES OF GEOMORPHOLOGY

While the cycle of erosion is primarily based on the sequential development of forms, the early work of Gilbert (1877), Dutton (1880) and their colleagues was followed by Penck's attempt (1925, 1953) to examine how hillslopes and channels evolved as a result of the relative importance of endogenetic and exogenetic forces. All denudational processes are essentially gravitational streams, gravitational forces being part of all sediment transport equations. All these gravitational streams tend downslope towards some kind of base level, which represents the termination of that particular kind of gravitational movement. For rivers there is a general base level in the sea, but for a blockstream the base level may be a decrease of gradient, for example the junction of a steep valley side with a gently sloping valley floor. Some of these levels regulating river action are associated with places where streams leave a tectonically uniform block, and enter an adjoining one which has undergone some different tectonic evolution. Renewed tectonic activity in this situation would alter local base level and so affect the evolution of channels and slopes. Changes in local base level are thus crucial factors in landform development.

Penck saw this change in local base level of denudation in terms of the relative efficiency of the exogenetic and endogenetic forces. If the general base level of denudation remains constant, decrease in the relative height of the land takes place more and more slowly. This slowing down comes about because, in the course of development, ever flatter and flatter slopes meet the zones of intersection; and on these denudation achieves less and less in successive units of time. The lowering of the zones of intersection depends solely upon the development of the slope units that meet there, and it is determined by whatever intensity of denudation is characteristic of these latter. But, Penck reminds us, the general base level of denudation is constant only as a special case.

There is a fundamental contrast between the course of development of landforms which is due to decrease in erosional intensity and is termed waning development (*absteigende Entwicklung*) and that which is bound up with increasing intensity of erosion as relative height increases during waxing development (*aufsteigende Entwicklung*). Tectonic stability would permit stable base levels of denudation, whereas tectonic activity would alter these base levels and give rise to waxing development. Thus Penck saw landscapes as a product of the dynamic balance between exogenetic and endogenetic processes. This balance has been established in such studies as Hjulström's investigation of the River Fyris (1935), which showed that while the land was rising 0·5 cm/yr, the river was only removing the equivalent of 0·0037 cm/yr, the area thus being in a state of waxing development, and Schumm's investigation of the disparity between rates of denudation and orogeny (1963c).

While Penck saw the equilibrium between tectonics and erosion, he did not emphasise the earlier observations by Gilbert (1880) that the rate of corrasion by a stream will depend on the character of its bed. Where the rock is hard corrasion will be less rapid than where it is soft, and there will result inequalities of grade. But so soon as there is inequality of grade there is inequality of velocity, and inequality of capacity for corrasion; and where hard rocks have produced declivities, there the capacity for corrasion will be increased. The differentiation will proceed until the capacity for corrasion is everywhere proportional to the resistance, and no further — that is until there is an equilibrium of action.

Assuming constant climate, uniform rate of uplift and constant base level, Penck claimed that the gradient of any river is a function of the rate of uplift with variations caused by differences of rock resistance, and that the rate of downcutting must everywhere equal the rate of upheaval (Simons, 1962). Penck pointed out that such a situation is largely theoretical for if upheaval becomes rapid, the water-rich lower course would erode more deeply than before, but the upper course would not cut so rapidly. A convex nick is formed in the long profile of the stream. This nick cuts back in accordance with the laws of headward erosion and forms the local base level for the upper reaches of the stream (Penck, 1925, quoted by Simons, 1962). Below the nick is a narrow steep course with convex valley slopes, above it a broader reach with concave slopes. Penck saw the development of hillslopes as a function of the balance, or lack of same, between uplift and denudation. When uplift and denudation are equal, straight slopes develop, but if uplift is accelerating, convex slopes develop. During waning development, when denudation is more rapid than uplift, concave hillslopes develop. Penck explained that the cliff face, or unit 4 of the 9-unit land surface model, would retreat parallel to itself, but the valley slope as a whole becomes concave and its gradient gradually declines (Tuan, 1958).

Many of the assumptions Penck made about processes and denudation rates have subsequently been found to be unwarranted. But Penck's model of landscape evolution has also suffered from oversimplification. It has been alleged that Penck suggested that when uplift and denudation are equal an equilibrium or time-independent landform develops. Such a situation will not occur; there is a fallacy in assuming that, because denudation rates are calculated as a uniform lowering of a land surface, denudation actually occurs in this manner. In fact, the forces of denudation are composed of two parts, hillslope erosion and stream channel erosion. Channel erosion may be rapid and in some cases approach the rate of uplift, present-day rates of orogeny generally greatly exceeding rates of denudation. However, hillslope erosion is much less effective than channel erosion. A theoretically perfect balance between rates of uplift and denudation will, therefore, manifest itself by channel incision and extension of the drainage pattern. Mescheryakov (1959), for example, attributes recent channel erosion in the south Russian steppe to contemporary uplift.

Dynamic equilibrium

In Chapter I it was suggested that the evolution of landforms could be studied in terms of the operation of a denudation system, an open system whose form and composition depend on the continuous import and export of materials and energy. As mentioned earlier in this chapter, Gilbert (1880) recognised that under constant conditions a stream will develop a characteristic, graded profile, but that if conditions alter, the profile will adjust to the new set of conditions. Strahler (1952b) described the graded profile as a 'time-independent steady state' which 'ensues when the energy developed through the descent of the water is entirely dissipated in overcoming resistance to shear within the fluid and against the channel boundary and to the movement of the bed load'.

From these notions of time-independent steady states in geomorphology, Hack (1960) developed the 'equilibrium concept of landscape', which asserts that the present topography of any area has developed by the nearly continuous down-wasting of a region. As the present landforms are adjusted to the net effects of past and present processes, most, if not all. traces of past changes which might have been expressed in the topography will have been destroyed. This argument hangs on the assumption that the form produced by a given process acting on any given rock is maintained as erosion proceeds. Hack (1960) argues that, in a given area, each kind of rock crossed by a given size of stream has its characteristic gradient, those on hard rocks being steeper than those on less resistant materials.

In an area of gently dipping sedimentary rocks, the slopes waste both horizontally and vertically; hence. they maintain the small overall gradient and form in each rock unit, both along streams and in interstream areas. For short periods of time certain components of the landscape, such as the outlying hills of the Hawkesbury Sandstone escarpment at Katoomba (Plate 3) may become out of balance. Thus when the caprock is removed from an area of soft rock, the soft rock is rapidly eroded and returns to a form of equilibrium. In terms of the landscape as a whole, this is part of the general process, and the evolution of the landscape is a continually changing pattern of similar forms. The escarpments and slopes adjust themselves as the outcrop patterns, which provide the geological framework, change.

Hack (1960) extends this argument to criticise the use of level surfaces in the landscape as evidence of past cycles of erosion by pointing out that because different rock types weather at different rates, two hills of different lithology but of the same elevation must necessarily be a different age, rather than erosional remnants of the same surface.

The primary problem involved in the application of Hack's hypothesis is to demonstrate conclusively that dynamic equilibrium has indeed been attained. In the previous chapter, the many cases of processes operating on present-day humid temperate landforms still adjusting to large deposits left over from Pleistocene morphogenesis under different climates were discussed. Many landforms adjust to such changes slowly; thus while they tend towards equilibrium with modern processes, they may in fact be far removed from a steady state today (Lustig, 1965). Morphology and process may be clearly interrelated, but the demonstration of dynamic equilibrium at any instance or time, even if conclusively proven, does not preclude the possibility that changes in both the intensities and types of processes may have occurred through time.

A secondary problem in Hack's theory is the assumption of more or less equal activity of the various processes over the whole landscape. This notion that 'denudation proceeds but is everywhere equal' is countered by Crickmay's hypothesis (1960, 1974) of unequal activity. Arguing that 'among all the real and imagined causes of a slope's retreat, only one has been established by observation, namely corrasion of its foot by water', Crickmay (1974) concludes that 'mobile, laterally corrasive rivers effect all the retreat of all the inland scarps that have ever retreated'. The presence or absence of such a river activity produces great discrepancies in erosional accomplishment, giving rise to areas where landforms are virtually unchanged for ages and to the rapid widening of valleys by lateral planation. While Crickmay draws attention to the need to recognise the spatial variability of the intensity of erosion, he does not completely invalidate the notion of adjustment between forces of resistance and forces of erosion, which is the basis of the dynamic equilibrium concept. Hack extended the concept to attempt to explain regional landform assemblages, but the major developments from the concepts have been in the application of probability theory and mathematical models to landform evolution.

PROBABILISTIC APPROACHES TO LANDFORM EVOLUTION

Strahler's advocacy of the use of mathematical statistics in geomorphology led to abundant empirical studies, particularly of the morphometric properties of drainage basins and slopes, and from them, to discussions of the most likely or most probable patterns of drainage network and slope evolution. Arguing that the relations between rocks and processes in space are the fundamental feature of the dynamic equilibrium approach, Scheidegger and Langbein (1966) find the concept close to the ergodic principle of statistical mechanics — the replacement of time averages by space averages and thus amenable to quantitative analysis.

A fundamental problem in studying natural systems, such as denudation systems, is that while the individual reactions, such as that between flowing water in a channel and a particle on the channel bed, act in a deterministic manner to produce a precisely predictable effect, so many individual reactions are involved that their combined result may be indistinguishable from the random (Scheidegger and Langbein, 1966). In studying denudation systems, a drainage basin may be defined, inputs in terms of solar radiation, precipitation, tectonic forces, atmospheric nutrients, and so on established, and outputs in terms of both runoff, loss of material, and changes of state evaluated. In landform evolution, the changes of state are the outputs of prime interest.

Devdarini and Greysukh (1969) suggest that in a number of cases it can be assumed, as a first approximation, that any state of the natural complex is determined only by the immediately preceding state and not by any earlier states. Such an assumption is probably relevant to the drainage basin, as it is the surface form and resistance at any one time which determines the work done by any hydrometeorological event. However, as not all the factors acting on inputs to the denudation system at any one time are known, the relationship between a state and the immediately preceding state will be stochastic: the transition from a given state to any other state would have a definite probability, written in the form of a stochastic matrix of transitions. The system can thus be modelled in the form of a Markov chain, or a probabilistic finite automaton. This chain would be nonhomogeneous: the matrix of transition probabilities will change over time in accordance with outside influences on the inputs of the system.

Markov processes are random in their occurrence but also exhibit an effect in which previous events influence, but do not rigidly control, subsequent events. The probability of a Markov process being in a given state at a particular time may be deduced from knowledge of the immediately preceding state. A Markov chain is a form of Markov process and may be regarded as a sequence or chain of discrete states in time (or space) in which the probability of the transition from one state to a given state in the next step in the chain depends on the previous state (Harbaugh and Bonham-Carter, 1970). The evolution of drainage networks has been simulated by Markov chain models.

Finite automata are mathematical models of real systems that process discrete information at discrete time intervals. The name 'automata' is used because such mathematical models were first built for modelling automatic engineering devices. Finite automata are those in which the number of possible states is finite. The study of the sequential evolution of landforms by finite automata was illustrated by Devdarini and Greysukh (1969) in terms of the probability of transformations in the sequence:

$$\text{rills} \rightarrow \text{furrows} \rightarrow \text{gullies} \rightarrow \text{ravines.}$$

While many probabilistic models of landform evolution have been developed (Scheidegger and Langbein, 1966; Devdarini, 1967a, b; Devdarini and Greysukh, 1969: Harbaugh and Bonham-Carter, 1970; Sprunt, 1972; Scheidegger, 1970), few really advance understanding of landform evolution, Sprunt (1972) arguing that few simulation models have yet shown more than was put into them. However, Scheidegger (1970), advocating the value of stochastic models, claims that as a diffusivity equation often results from the setting up of stochastic models, a general analogy between landscape variable and thermodynamic variables offers much to the theory of landform evolution.

In two dimensions (Cartesian co-ordinates x, y) the thermodynamic field is described by the temperature $T(x, y)$ and the quantity of heat Q_*. The landform field is described by the height $h(x, y)$ of the land above some base line and the mass M_*. The following analogues thus exist (Leopold and Langbein, 1962):

$$T \quad - \quad h$$
$$dQ_* \quad - \quad dM_*.$$

Corresponding entropies U in thermodynamics and landscape evolution may be defined:

$$dU = dQ_*/T - dM_*/h$$

Developing this thermodynamic analogy, Scheidegger (1970), establishing the similarity between the diffusivity equation for temperature arising from heat conduction in a thermodynamic system and for height using from the diffusion of height during degradation of a land mass, shows that:

$$\frac{\partial h}{\partial t} = D\left[\frac{\partial^2 h}{\partial x^2} + \frac{\partial^2 h}{\partial y^2}\right]$$

The application of these ideas to the 'decay' of a mountain range yields a succession of profiles whose shape is much that of the Gaussian normal distribution curve, showing loss of height in the central and highest area of the range and the accumulation of debris on the outer margins of the range (Fig. 52).

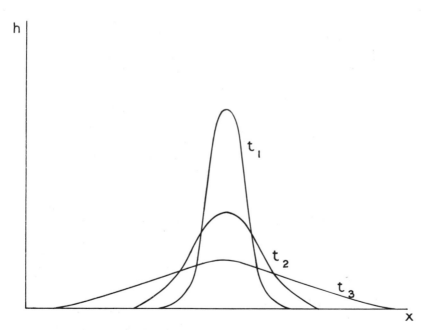

52 *Theoretical decay of a mountain range according to the diffusivity equation*

Harbaugh and Bonham-Carter (1970) report Pollack's use of a modified form of the diffusion equation in modelling the development of deep rock-cut, canyon-like valleys. The representation of various strata in terms of their erosional characteristics enabled valley cross-profile development to be simulated. Several pulses of development occur in the model, with rapid deepening when a non-resistant bed is encountered and then valley widening in which the profile 'diffuses' outward from the stream site. Such a model shows changes in form with time, topographic features evolving and later being eliminated, thus suggesting the feasibility of developing stream erosion models in three dimensions.

The next step would be to start with an irregular surface and suggest the way in which such a surface would evolve due to its initial effects on directions and rates of runoff and sediment transport. The models described in the literature are largely linear and imply uniform initial surfaces. Such assumptions would be justified in much the same way as Davis justified the assumption of initial rapid uplift in the 'ideal' cycle of erosion. The three-dimensional, irregular surface possibility is inherent in Sprunt's work (1972) and is readily capable of further development.

One problem with the application of these ideas has been the real lack of adequate information on which to base simulation

TABLE 18 **Factors controlling drainage density**
(after Strahler, 1964)

Symbol	Term	Dimensional quality
Dd	Drainage density	Length divided by area
Qr	Runoff intensity	Volume rate of flow per unit area of surface
K	Erosion-proportionality factor	Mass rate of removal per unit area divided by force per unit area
h	Relief	Length
ρ	Density of fluid	Mass per unit volume
μ	Dynamic viscosity of fluid	Mass per unit length per unit time
g	Acceleration due to gravity	Distance per unit time per unit time

runs and to check the validity of models. While modern remote sensing and digitising equipment should make the acquisition of such data fairly easy, there remains a considerable gulf between field observation and computer simulation. Not surprisingly, readily two-dimensional data on the topology of drainage networks have been the subject of more computer simulation than most other landform attributes.

Functional relationships between landform and process variables

While general probability models lack precision, the controls of drainage density can be specified with reasonable precision. Strahler (1964) uses drainage density (Dd) as the dependent variable in developing an equation relating the horizontal scale of landform units to a series of independent or controlling variables (Table 18):

$$Dd = f(Qr, K, h, \rho, \mu, g)$$

The variables in this equation may be grouped into the functional relationship:

$$f^1(Dd, Qr, K, h, \rho, \mu, g) = 0$$

The seven variables in this function may be reduced to four through application of the pi theorem. If any variable a depends upon the independent variables $a_2, a_3, \ldots a_n$, and upon no others, the relationship may be written as:

$$a = f(a_2, a_3, \ldots a_n)$$

The pi theorem states that, if all these n variables may be described with m fundamental dimensional units, they may then be grouped into $n - m$ dimensionless parameters, or terms:

$$f(\pi, \pi_2 \ \pi_3 \ldots \pi_{n-m}) = 0$$

In each term there will be $m + 1$ variables, only one of which need be changed from term to term. Solution of the four pi terms in the drainage density function yields a function of four dimensionless groups:

$$\phi \, (\text{h Dd, Qr K,} \ \frac{\text{Qr}\rho\text{h}}{\mu}, \frac{\text{Qr}^2}{\text{hg}}) = 0$$

Solving for drainage density gives:

$$\text{Dd} = \frac{1}{\text{h}} \, \text{f} \, (\text{Qrk,} \ \frac{\text{Qr}\rho\text{h}}{\mu} \frac{\text{Qr}^2}{\text{hg}})$$

The term hDd is the ruggedness number, a morphometric parameter expressing the essential geometrical characteristics of the drainage system. The second term, QrK, is the Horton number, expressing the relative intensity of the erosion process in the drainage basin. The third term, Qr ph/μ is a form of Reynolds number, in which Q takes the place of velocity and h the characteristic length. The fourth term, Qr^2/hg, is a form of Froude number.

This reduction of the seven variables into four dimensionless groups focuses attention on dynamic relationships, simplifies the design of field measurements programs, and establishes a way of making comparisons between mathematical models and natural drainage systems. Using these relationships the effects of changes in runoff intensity, erodibility of the land surface, or of any factors affecting these variables can be assessed. Any tectonic activity would have to be incorporated in terms of intensity of erosion, but essentially these dynamic relationships operate over time spans of hundreds rather than millions of years.

Three-dimensional models of drainage basin evolution

Both the Scheidegger mountain range decay model and Pollack's canyon-like valley evolution model discussed earlier examine profiles through surfaces, while most drainage network models are two-dimensional plan-view situations. Strahler's pi theorem work has had little application, but form and process have been brought together in models of drainage basin evolution by Smith and Bretherton (1972). Two mathematical models embodying the following assumptions are put forward:

1. A drainage basin may be represented by a mathematical surface.
2. The principle of the conservation of mass is applicable to this surface.

3. The transport of sediment at any point of the surface may be adequately described by some function of the local slope and the local discharge of water.

The basic issue is again the question of how an initial surface changes through time, and assumptions about homogeneous lithology, steady rainfall rate and no infiltration or evaporation are necessary. The smooth surface model depends on sediment transport being described as

$$Q_T = kQ^j S$$

With all the above assumptions the model is consistent with Horton's critical belt of no erosion, described in Chapter VI, and with the way in which badlands often have small unchannelled convex upper slopes and unchannelled straight slopes leading either directly to channels or small, seasonally channelled convex slopes (Schumm, 1956a, 1956b). The model thus reproduces units 2, 3, 5 and 6 of the 9-unit land surface model, those units where surface wash and transport processes are dominant.

The second model attempts to cope with the lateral migration of river channels considered so important by Crickmay (1960, 1974). Lateral migration is taken to be away from the slope supplying the greatest volume of sediment. Problems arose in considering how lateral migration leads to streams eventually joining together, in a captive process, analogous to down-dip breaching (Pitty, 1965) and the evolution of conjoint floodplains (panplains) (Crickmay, 1933). However, this model does throw light on the valley widening process noted in the previous chapter with reference to the change from the *Kerbtal* to the *Sohlenkerbtal* in rugged humid tropical terrain.

These two models providing information on initial channel cutting and channel shifting indicate how some of the fundamental issues of humid landform evolution may be resolved. As with their precedents, they accord better with the dynamic equilibrium concept than with longer term theories. However, a marriage of this type of approach with some of the geophysical, structural and sedimentary models developed in geology (Harbaugh and Bonham-Carter, 1970; Krumbein and Graybill, 1965) offers great potential for longer term landform evolution investigation.

While the developments in mathematical modelling have not yet led to major additions to theory, they have had the salutary effect of asking precisely how geomorphic processes work and what the precise relationship between form and process are. Their association with shorter time spans is inevitable at this stage of their development. As time spans increase, fewer and fewer variables may be considered constant. Climatic change, the evolution of life forms, and the development of the oceans and continents all affect humid landform genesis even down to the scale of comparisons between one part of a land area such as Australia with another. Any analysis of the history of scenery requires consideration of continental drift, changes in organic species and climate. Events of this order involve great feedback loops within the denudation system, but also imply that steady progress of landform evolution towards a final form is unlikely, and that erosion surfaces are probably of composite origin and not the equivalent of the peneplain produced by 'normal' erosion or of any other single planation process.

To bring together the form-based approach of the Davisian school and the dynamic approaches to geomorphology, a distinction has to be made between the millions of years required for peneplanation, the tens of thousands of years of sea level oscillations and climatic change, and the annual, seasonal and diurnal fluctuations of the hydrometeorological variables of the denudation system. If the age of the earth is considered to be approximately 4000 million years, and this age is taken to be time span 1, and then successively smaller time spans are to be taken by powers of ten, time (T) 2 would be 400 million years, T3 40 million and so on to T10 which is 4 years, T11 which is 20 weeks and T12 which is 15 days. In the longer time spans, up to T4, the tectonic and denudation factors of the geographical cycle prevail and the Davisian model may be applied. In the middle range time spans from T5 to T8 climatic equilibria are relevant. In the shorter time spans minor adjustments of the system occur in response to the variability of hydrometeorological factors. Therefore the significance of the factors that affect landform evolution and the applicability of conceptual models can be seen to vary with time.

In humid landform investigation it is appropriate to consider these factors as drainage basin variables. The variables which, over a given time span, are constant and determine the way in which the drainage basin hydrological and denudation systems operates may be termed independent variables, whereas those which vary as a consequence of the operation of the system through that time span, are termed dependent variables. Table 19 shows how drainage basin variables which are dependent in the cyclic (T4) time are independent variables during shorter time spans.

TABLE 19 **The status of drainage basin variables during time spans of decreasing duration**
(after Schumm and Lichty, 1965)

Drainage basin variables	Status of variables during designed time spans		
	T4 (cyclic)	T10 (graded)	T12 (steady)
1 Time	Independent	Not relevant	Not relevant
2 Initial relief	"	"	"
3 Geology (lithology, structure)	"	Independent	Independent
4 Climate	"	"	"
5 Vegetation (type and density)	Dependent	"	"
6 Relief or volume of system above base level	"	"	"
7 Hydrology (runoff and sediment yield per unit area within system)	"	"	"
8 Drainage network morphology	"	Dependent	"
9 Hillslope morphology	"	"	"
10 Hydrology (discharge of water and sediment from system)	"	"	Dependent

As mentioned earlier, geomorphology is a study in four dimensions, length, breadth, depth and time. Questions of shape, form, spatial interrelationships and rates of change are constantly raised in the investigation of humid landforms. The understanding of the delicate adjustments occurring in the present-day landscape yields as much satisfaction to the student of the face of the earth as the unravelling of the varied evidence of landform history and the cross-correlation of one dated horizon or feature with another. The scenic delights of water tumbling over the caprock of a high fall and crashing into the bank of a gorge 300 m below are only heightened by an appreciation of the significance of that feature for the history of the surrounding landscape. The knowledge of the vital geomorphic contrast a change of elevation of 1 m implied in a delta or coastal plain is equally rewarding and leads to a wider appreciation of the ever changing denudation system. Humid landforms, in particular, with their cover of soils and vegetation, embrace the major inhabited portions of the earth and have been highly susceptible to modification by the direct and indirect action of man. Knowledge of the morphological activity of rivers, the denudation of hillslopes and the formation of depositional landforms is vital to successful use of the land. Man's actions cannot be understood without some comprehension of the processes described in this book. The record of man's achievements and destructive works as a geomorphic agent is, however, another story.

APPENDIX

NOTATION

A	cross-sectional area of flow
C	Chézy discharge coefficient
C_*	cohesion or cohesiveness
D	particle size diameter
D_{84}	particle diameter equal to or larger than 84 per cent of the particles in the sample
Dd	drainage density
d	depth of flow
E	specific energy
E_*	kinetic energy in J
F_B	bed factor
F_S	side factor
f	Darcy-Weisbach coefficient
g	acceleration due to gravity
H	weighted mean percent silt-clay in the channel perimeter
h	height
I	rainfall intensity in mm hr^{-1}
J	Joule
j	an exponent
K	erosion proportionality factor
k	constant coefficient
L	a constant
M_L	meander length

239

M_W	meander width
M_*	mass
n	roughness coefficient
P	sinuosity (ratio of channel length to valley length)
p	a constant
P_*	length of wetted perimeter in a cross-section
Q	total discharge
Q_B	bed material load
Q_s	suspended sediment load
Q_T	total sediment load
Q_r	runoff intensity
Q_*	quantity of heat
R	hydraulic radius
Re	Reynolds number
r	erosion index of Wischmeier and Smith
S	channel gradient, or sine of channel gradient
S_*	shear strength
T	temperature
U	entropy
V	mean flow velocity
v	kinematic viscosity
W	width of channel flow
w_c	channel width
σ	pressure normal to the shear plane
μ	dynamic viscosity
o	density of fluid
ϕ	angle of shearing resistance or internal friction

GLOSSARY

Abrasion	The mechanical process of erosion by the friction of rock particles in transit.
Absorption	The process by which substances penetrate the surfaces of solids.
Adsorption	The processes by which substances adhere to surfaces of solids. The adsorbed substances may include ions, molecules, or colloidal particles.
Aggradation	The filling in of low areas by the deposition of sediment.
Anastomosing Channels	A system of interlaced channels formed by consequent runoff passing around pre-existing obstructions of alluvium or bedrock.
Angle of Repose	The angle between the horizontal and the slope of a heap produced by pouring dry sand from a small height; this angle is approximately equal to the angle of internal friction, the angle at which grains resting on each other will remain in position.
Attrition	The wear of rock particles in transit.
Bankfull Discharge	The discharge at the stage when the elevation of the water surface of a stream is at channel capacity, just level with the top of the stream bank.
Base Flow	The dry period, low flow of a river derived from groundwater discharge.
Bauxite	The principal ore of aluminium; composed of hydrated oxide containing various impurities.
Bed Load	Material transported by rolling, sliding or jumping (saltating) close to the channel bed.
Bench Mark	A cement marker and plate which records the

exact geographical position and elevation of a point. Also used for any reference point, such as a blaze on a tree, used for surveying.

Boreal

Pertaining to the north of the northern hemisphere, or to climatic zones with a definite winter that experiences snow and a short summer that is generally hot, and characterised by a large annual range of temperature.

Braiding

The process by which a river divides into many interlocking channels within a single floodplain.

Boundary Shear Stress

The shear stress on the wall of a pipe or perimeter of a channel.

Calibre

The size of rock particles involved in fluvial denudation.

Cambering

The process by which rocks at or close to the surface acquire an arched or curved structure due to gravitational sagging towards topographically lower areas.

Capillary Fringe

The zone immediately above the water table in which water rises above the water table in a mass of irregular capillary tubes.

Capillary Tension

The stress holding water equilibrium by surface tension in the pore spaces between particles immediately above the water table in soils or regolith material.

Carbon^{14} Dating

Carbon-bearing materials which are formed in equilibrium with atmospheric carbon acquire during their formation small amounts of the cosmic-ray-produced C^{14} isotope. When the exchange with the atmospheric carbon comes to an end, the radioactive decay of the C^{14} provides a method for determining the period elapsed since the cessation of the exchange and hence, by inference, the age of the deposit in which the material was found.

Chlorophyll

Complex organic molecule that is a necessary catalyst in photosynthesis.

Clay Mineral

Hydrous iron and aluminium silicate minerals with various other absorbed and adsorbed ions.

Climatic Geomorphology	The study of the influence of climate on the evolution of landforms.
Cohesiveness	A measure of the strength of the soil under undrained conditions.
Colloid	A material so finely divided that interfacial physical and chemical properties predominate. A gluelike substance whose particles (molecules or aggregates of molecules) when dispersed in a solvent remain uniformly suspended and do not form a true solution.
Colluvium	Weathered debris including soil which has moved downslope under the influence of gravity.
Corestone	A more or less intact, rounded remnant of a joint block isolated by weathering of the surrounding rock.
Corrasion	The wearing of a rock surface by moving rock particles.
Critical Depth	The point on the specific energy curve where specific energy is least.
Cutan	Deposit of concentration of small particles and colloidal material on the surfaces of peds, pores, sand grains or stones within a soil or weathering profile.
Denudation	The removal of detrital material through erosion and mass wastage on the earth's surface.
Denudation System	The sum of the interacting processes which result in the wearing away or the progressive lowering of the earth's surface by various natural agencies.
Diagenesis	The changes which occur in a sediment following deposition, accumulation and burial.
Discharge	The rate of flow at a given time expressed as volume per unit of time.
Dominant Discharge	That discharge of a natural channel which determines the characteristics and principal dimensions of the channel, It depends on the sediment characteristics, the relationship between maxi-

mum and mean discharge, duration of flow, and flood frequency.

Duricrust

Ground surface mineral incrustation formed by mobilisation and reprecipitation of ions.

Dynamic Equilibrium

A least-work, equal area energy expenditure condition for a system characterised by imperceptible, short-term physical changes.

Ecosystem

A given abiotic physico-chemical environment and its particular biotic assemblage of plants, animals and microbes. Ecosystems are real — like a pond, a field, a forest, an ocean, or even an aquarium — but they are also abstract in the sense of being conceptual schemes developed from a knowledge of real systems.

Epiphyte

A plant which uses other plants for support (for example, some mosses and orchids).

Erodibility

The degree to which a material is likely to be affected by rain, water or wind processes leading to its disintegration and removal.

Erosion

Any or all of the processes that loosen and remove earth or rock material.

Evaporation

The taking up of moisture from a surface by under-saturated air.

Evaporation Crust

The hard crust formed by the concentration of salts on or just below the ground surface as a result of loss of water to the atmosphere by evaporation.

Evapotranspiration

The combined processes of evaporation and transpiration by plants as a result of which water is lost from a vegetated land surface to the atmosphere.

Fabric

The detailed organisation, structure and intergranular relationships of deposit or soil.

Fall-making Rock

The resistant rock which forms the upper lip of a waterfall.

Fen

Marshy, swampland vegetation of grasses, sedges, cattail etc. growing on peat over wet subsoil.

Ferricrete	A form of duricrust with iron and usually some silica.
Floodplain	A flat plain adjacent to a river and composed of sediments deposited by that river during floods.
Fluvial System	A set of processes of environmental change regulated by running water over and within the earth's crust.
Geochemical Cycle	The translocation of chemical elements through the sequence of rock weathering, erosion, deposition, compaction, hardening, metamorphism and remelting to form new rock again.
Graded Profile	The longitudinal profile of a river in which all irregularities due to lithology, tectonics or change in altitude of the river mouth have been eliminated. Defined by Mackin (1948) as one 'in which, over a period of years, slope and channel characteristics are delicately adjusted to provide, with the available discharge, just the velocity required for the transportation of the (sediment) load supplied from the drainage basin'.
Granulometry	The establishment of geometric parameters, length, breadth, rounding, etc., to describe rock fragments, sand grains and other sedimentary particles.
Glaciofluvial	Concerned with the activity of, and deposition by, meltwater at the margin of, ·beneath, and even on, glaciers.
Greywacke	A dark (usually grey or greenish grey, sometimes black) and very hard, tough, and firmly indurated, coarse-grained sandstone consisting of poorly sorted, angular to subangular grains of quartz and feldspar with abundant small, dark rock and mineral fragments embedded in a clayey matrix with shale and slate. A term widely and loosely used for hard, dark, clayey sandstones interbedded with shales or slates.
G-scale	A scale for areal units based on the area of the earth.
Hydration	The chemical combination of water and another substance, a process in which water is added to

the substance involved.

Hydraulic Geometry

The description, at a given cross-section of a river channel, of the graphical relationships among plots of hydraulic characteristics (such as width, depth, velocity, channel slope, roughness, and bed particle size, all of which help to determine the shape of a natural channel) as simple power functions of river discharge.

Hydraulic Radius

The cross-sectional area of a channel divided by the length of the wetted perimeter of the cross-section.

Hydrological Cycle
(sometimes termed water cycle)

The course taken by water in moving from the oceans to the land via evaporation and precipitation and returning via streamflow, including via soil water and groundwater. The subsystem of the global energy system that regulates the flow of energy through the heat-exchange property of water.

Hydrolysis

The process by which a hydrogen ion enters the atomic structure of a mineral in exchange for a cation.

Hydrostatic Pressure

The pressure exerted by the water at any given point in a body of water at rest. The hydrostatic pressure of groundwater is generally due to the weight of water at higher levels in the zone of saturation.

Hydrothermal

Of or pertaining to heated water, to the action of heated water, or to the products of the action of heated water, such as a mineral deposit precipitated from a hot aqueous solution, with or without demonstrable association with igneous processes.

Interception

The proportion of rainfall held on the foliage of plants.

Infiltration

The penetration of the ground surface by water.

Interfluve

The area between two rivers; the area about the divide separating one set of valley-side slopes from another.

Krasnozem

A red or reddish-coloured soil developed under

humid conditions with a profile which grades steadily from the surface through weathered rock to unweathered rock.

Kurtosis A measure of the peakedness of a frequency distribution, e.g. a measure of the concentration of sediment particles about the median diameter.

Lag Gravel Coarse surficial sediment residue resulting from loss of fine material by sheetwash and aeolian deflation.

Lateral Migration The shift of river channels across a floodplain or other surface by the corrosion of banks at the outer sides of stream bends and associated collapse of stream margins.

Laterisation The soil forming process characteristic of wet but well-drained sites, particularly in warm climates, which leads to the accumulation of sesquioxides of iron and to relative removal of silica and the development of soil profiles which grade from the surface to the unweathered rock without distinct horizons.

Leaching Removal of constituents of rock material by the solvent action of percolating water.

Limit of Liquidity (liquid limit) The water content at which a trapezoidal groove of specified shape, cut in moist soil held in a special cup, is closed after 25 taps on a hard rubber plate.

Limit of Plasticity (plastic limit) The water content at which the soil begins to break apart and crumble when rolled by hand into threads 3·2 mm (1/8 inch) in diameter.

Lithology The study of the character of rocks.

Longitudinal Profile The graph of altitude of a channel bed plotted against distance from either source or mouth.

Lower Flow Regime Subcritical turbulent flow.

Magma A mass of molten rock beneath the earth's surface.

Markov Chain A statistical model developed to incorporate a notion of dependency, each event in the sequence depending on the event immediately prior to it.

	In geomorphology, each state of the land surface thus depends on the immediately previous state.
Mass Failure	The sudden collapse of a slope in a complex of slips and slumps.
Metabolism	The processes of synthesis or destruction of protoplasm in living cells and, thus, higher organisms.
Micro-morphology	The study of undisturbed thin sections of soils under the microscope to examine the details of soil fabrics.
Morphoclimatic Region	An area with a distinct climatic regime, within which the intensity and relative significance of the various geomorphic processes are essentially uniform.
Morphodynamic	Concerning the changing of landforms through time.
Morphogenic Environment	A set of conditions under which a certain group of agencies and processes interact to produce an assemblage of landforms.
Morphological Map	A pictorial representation to scale on a flat surface of the character, and sometimes origin, of landforms.
Normal Erosion	Subaerial erosion by running water, rain, and certain physical and organic weathering processes. The term, used originally for stream erosion in a temperate climate, is open to criticism because erosion as found in temperate areas may in fact be 'abnormal' (especially in regard to past geologic conditions) or because one mode of erosion is just as 'normal' as another.
Overland Flow	Movement of water over the ground surface, as unconcentrated wash, diffuse flow or linear flow.
Oxidation Potential	A quantitative measure for the energy change involved in adding or removing electrons to an element in a particular oxidation state; a measure of the stability of an element with reference to the standard reaction

$$H^2 = 2H^+ + 2e \ (e = \text{electron})$$

of which the oxidation potential is arbitrarily fixed at 0·00 volt.

Palingenesis	Formation of a new magma by the melting of pre-existing rock *in situ*.
Palynology	The study of spores and pollen grains, especially their presence in sediments of lakes and bogs.
Partial Area	The sectors of a catchment area which, in a given rainstorm, become saturated and contribute to streamflow.
Ped	A soil aggregate with distinct edges visible in an undisturbed thin section.
Perched Water Table	A groundwater body situated above the main one of a region on an impervious rock layer.
Percoline	Line of diffuse subsurface water movement above the head of a permanent channel.
Periglacial	Nonglacial processes and features of cold climates regardless of age and of any proximity to glaciers. The environment dominated by significant frost action and snow-free ground for part of the year.
Permeability	The property of rock, sediments or soils which permits liquids to pass through them.
Petrology	The branch of geology dealing with the origin, occurrence, structure and history of rocks, especially igneous and metamorphic rocks.
pH	The negative logarithm of the hydrogen ion concentration in an aqueous solution.
Photosynthesis	The manufacture of starches and sugars by plants using air, water, sunlight and chlorophyll.
Piping	Process creating subsurface tunnel or pipe in the weathering profile above a stream or gully head.
Podsolisation	A soil forming process resulting from leaching by acid soil solutions.
Point Bar	An accumulation of relatively coarse bed material

on the concave side of the talweg adjacent to a pool.

Pool

A topographically low area in a channel produced by snow and generally containing relatively fine bed material.

Pore Pressure

The stress transmitted through the fluid that fills the voids between particles of a soil or rock mass; e.g. that part of the total normal stress in a saturated soil due to the presence of interstitial water.

Process-response Model

A representation of a natural system in which changes in processes (flows and stress forces) produce changes to landforms which in turn lead to changes in the processes.

Regime

The condition when a river displays virtually constant channel form and gradient over extended reaches.

Regolith

The unconsolidated debris mantling bedrock.

Rejuvenation

Accentuated stream incision, often attributed to uplift, or to lowering of the river mouth.

Relative Roughness

The ratio of depth to the size of the roughness elements in a channel.

Remote Sensing

The collection of data about a surface or area using distant sensors for electromagnetic radiation of any wavelength, in aircraft, satellites, or balloons.

Residual Shear Strength

Under continuing strain, a dense soil reaches a peak strength followed by a loss of strength (and increase in void ratio). The strength remaining after large strain is termed the *residual* strength.

Riffle

A topographic high area of a channel, produced by the lobate accumulation of relatively coarse bed material.

Rill

A narrow, linear groove or depression developed by water flowing over unconsolidated material on slopes.

Ripple

A small elongated low ridge produced by water

(or air) flowing over unconsolidated sediment.

Roughness Coefficient	A numerical expression of the way in which irregularities and obstructions on the channel perimeter cause loss of energy by the flowing water.
Scree	Loose rock debris stream down a slope.
Selva	The tropical rain forest environment, characterised by a dense forest of broadleaved evergreen trees and high humidity. The term is also used by Garner (1974) to mean: Topography entirely reduced to slopes between narrow ridges and channel floors at the bottoms of V-shaped valleys, generally under humid conditions unless the materials being eroded are fine-grained and poorly consolidated.
Shear Line	A line along which differential movement has occurred as a result of stresses which caused contiguous parts of a body to slide relatively to each other in a direction parallel to their plane of contact.
Shear Strength	The shear stress of failure; the ability of a material to withstand a stress acting in the surface of the plane of the material.
Shearing Resistance	A tangential stress caused by fluid viscosity and taking place along the boundary of a flow in the tangential direction of local motion; the force which has to be overcome before a given particle can be taken into motion by a flowing medium.
Sheeting	The splitting away of thick rock layers, roughly parallel to the surface along fractures commonly due to unloading stresses.
Shrinkage Limit	The water content at which a soil reaches its theoretical minimum volume as it dries out from a saturated condition.
Sinuosity	The ratio of talweg length to valley length or to flow distance in absence of a valley.
Skewness	The degree to which a sample distribution or statistical population has a single mode like the

	normal distribution, but is asymmetrical; a distribution highly affected by extreme values.
Soil Moisture	Water in the soil.
Solar Radiation	Radiant energy from the sun.
Solifluction	Slow, downslope movement of water-saturated regolith, often in association with frost action or wetting and drying.
Solubility	The equilibrium concentration of a solute in a solution saturated with respect to that solute at a given temperature and pressure.
Solute	A dissolved substance.
Specific Energy	Total head above the floor of a channel, head being the hydrostatic pressure of a liquid divided by the specific weight of the liquid.
Spheroidal Scaling	Weathering propagated inward from fracture surfaces in jointed bedrock to produce rounded residual centres.
Stochastic	Concerned with the calculation of probabilities of changing to one state as opposed to another. Stochastic models are used to forecast likely future states of landforms on the assumption of step by step evolution dependent on the probability of development taking one of a number of possible directions at each step.
Stemflow	The proportion of rainfall flowing along branches and down the stems of plants and reaching the ground surface.
Stream Order	The ranking of a stream according to the number of tributaries feeding into it.
Stress	A force per unit of area. A stress applied to a plane surface of a solid can be resolved into two components: one perpendicular (normal) to the plane known as the *normal* stress, and one acting in the surface of the plane known as the *shear* stress.
Subcritical Flow	Flow at depths greater than the critical depth.

Supercritical Flow	Flow at depths less than the critical depth.
Suspended Load	Solid particles carried in such a way that the particles are surrounded by water.
Swale	The depression between bar-like forms or successive levees on a flood plain; the depression between successive dunes on a beach or in a desert.
Swirlhole	A circular depression scoured in a stream bed by rock fragments carried in a rotating current.
Talus	Loose rock debris at the base of a slope.
Talweg	The line running through the deepest parts of a river channel.
Tectonic Warping	Deformation of the land surface over large areas due to differential movement within the crust, particularly that caused by subcrustal movements.
Tension Crack	A fracture developed perpendicular to the direction of greatest stress and parallel to the direction of compression, caused by tensile stress.
Terrace	Elevated portions of alluvial fills, rock-cut benches, or other planar features along a stream valley.
Threshold Slope Angle	The maximum at which a slope is stable before failure occurs.
Throughfall	The proportion of rainfall falling between the leaves and branches of plants and reaching the ground surface.
Throughflow	The flow of subsurface water through voids in the regolith downslope towards channels or the water table.
Tor	A bare rock mass surmounted and surrounded by blocks and boulders.
Unloading	The process of strain release when a rock formed at high temperature and pressure beneath the earth's surface becomes exposed to subaerial

weathering through the removal of the over-burden.

Upper Flow Regime   Supercritical turbulent flow.

Vigil Network   A basic reference network of precise measurements of landform features for resurvey every few years to establish rates of landform change under various environmental conditions.

Waste Mantle   The layer of broken rock material, *in situ* or translocated, covering the unweathered and unbroken parent rock of a slope.

Water Balance   The budget for the operation of the hydrological cycle that states precipitation minus evapotranspiration and groundwater discharge equals runoff.

Weathering   The breakdown of rocks caused by the action of atmospheric agents.

Weathering Profile   The sequence of layers, including soil from the ground surface to the weathering front, or basal surface which separates the decomposed or partly decomposed rock from the unweathered material beneath.

Zone of Saturation   The area below the water table in rocks or soils where all voids are completely filled with water.

BIBLIOGRAPHY

Ackers, P. and Charlton, F. G. (1970). Meander geometry arising from varying flows. *J. Hydrol.*, **11**, 230–52.

Ahmad, F. (1973). Have there been major changes in the Earth's axis of rotation through time? In Tarling, D. H. and Runcorn, S. R. (eds.), *Implications of Continental Drift to the Earth Sciences*, Volume 1. Academic Press, London, 487–501.

Allen, J.R.L. (1970). *Physical Processes of Sedimentation: An Introduction*. Allen and Unwin, London.

Armstrong, P. H. (1972). Bristlecone pines tell an 8000-year story. *Geogrl. Mag. London.*, **44**, 637–9.

Arnett, R. R. (1971). Slope form and geomorphological process: an Australian example. *Inst. Br. Geogr. spec. Publs.*, **3**, 81–92.

Axelsson, V. (1967). The Laitaure delta. A study of deltaic morphology and processes. *Geogr. Annlr*, **49A**, 1–127.

Bagnold, R. A. (1941). *The Physics of Blown Sand and Desert Dunes*. Methuen, London.

— (1956). The flow of cohesionless grains in fluids. *Phil. Trans. R. Soc. Am.*, **249**, 235–97.

Bakker, A. J. and Van Wijk, C. L. (1940). Infiltration and runoff under various conditions on Java.

Bakker, J. P. and Le Heux, J.W.N. (1946). Projective-geometric treatment of O. Lehmann's theory on the transformation of steep mountain slopes. *Proc. K. ned. Akad. Wet.*, **49**, 533–47.

— and — (1947). Theory on central rectilinear recession of slopes. *Proc. K. ned. Akad. Wet.*, **50**, 959–66 and 1154–62.

— and Strahler, A. N. (1956). Report on quantitative treatment of slope recession problems. *Report of the Commission for the Study of Slopes, International Geographical Union*, 1–12.

Baltzer, F. (1972). Quelques effets sédimentologiques du Cyclone Brenda dans la plaine alluviale de la Dumbea (Côte ouest de la Nouvelle Calédonie). *Révue Géomorph. dyn.*, **21**, 97–116.

Barnes, H. H. (1967). Roughness characteristics of natural channels. *Wat. Supply Pap. Wash.*, 1849.

Barrell, J. (1912). Criteria for the recognition of ancient delta deposits. *Bull. geol. Soc. Am.*, **23**, 377–446.

Barshad, I. (1972). Weathering — chemical. In Fairbridge, R. W. (ed.), *The Encyclopedia of Geochemistry and Environmental Sciences* (Encyclopedia of Earth Sciences Series Vol. IVA), Van Nostrand Reinhold, New York, 1264–9.

Baulig, H. (1940). Le profil d'équilibre des versants. *Annls Géogr.*, **49**, 81–97.

Bayly, I.A.E. and Williams, W. D. (1973). *Inland Waters and their Ecology*. Longman, Camberwell.

Behrmann, W. (1921). Die Oberflächenformen in den feucht-warmen Tropen. *Z. Ges. Erdk. Berl.*, **56**, 44–60.

— (1933). Morphologie der Erdöberfläche. In Klute, F. (ed.), *Handbuch der geographischen Wissenschaft*. Akademische Verlagsgesellschaft Athenaion, Wildpark-Potsdam.

Berry, L. and Ruxton, B. P. (1961). Mass movement and landform in New Zealand and Hong Kong. *Trans. R. Soc. N.Z.*, **88**, 623–9.

Bik, M.J.J. (1967). Structural geomorphology and morphoclimatic zonation in the Central Highlands, Australian New Guinea. In Jennings, J. N. and Mabbutt, J. A. (eds.), *Landform Studies from Australia and New Guinea*. A.N.U. Press, Canberra, 26–47.

Bird, E.C.F. (1968). *Coasts*. A.N.U. Press, Canberra.

Birot, P. (1966). *General Physical Geography*. Translated by Margaret Ledésert. Harrap, London.

Bishop, W. W. (1967). The Lake Albert basin. *Geogrl J.*, **133**, 469–80.

— and Trendall, A. F. (1967). Erosion-surfaces, tectonics and volcanic activity in Uganda. *Q. Jl geol. Soc. Lond.*, **122**, 285–420.

Blackwelder, E. (1925). Exfoliation as a phase of rock weathering. *J. Geol.*, **33**, 793–806.

— (1933). The insolation hypothesis of rock weathering. *Am. J. Sci.*, **226**, 97–113.

Blake, D. H. and Paijmans, K. (1973). Landform types of eastern Papua and their associated characteristics. *Land Res. Ser. CSIRO Aust.*, **32**, 11–21.

Blench, T. (1957). *Regime Behaviour of Canals and Rivers*. Butterworths, London.

— (1961). Hydraulics of canals and rivers of mobile boundary. In *Butterworth's Civil Engineering and Reference Book*. (2nd ed.), Butterworth, London.

— (1973). Regime problems of rivers formed in sediment. In Shen, H. W. (ed.), *Environmental Impact on Rivers (River Mechanics III)*. The Editor, Fort Collins, 5–1 to 5–33.

Bos, R.H.G., Jungerius, P. D., and Wiggers, A. J. (1971). Solifluction and colluviation on the ice-pushed ridges of Uelsen, Kreis Grafschaft Bentheim, Germany. *Geologie Mijnb.*, **50**, 751–4.

Bowman, H. N. (1972). Natural slope stability in the City of Greater Wollongong. *Rec. geol. Surv. N.S.W.*, **14**, 159–222.

Brade-Birks, G. (1946). *Good Soil*. Edinburgh University Press, London.

Bray, J. R. and Gorham, E. (1964). Litter production in the forests of the world. *Adv. ecol. Res.*, **2**, 101–57.

Brewer, R. (1972). The basis of interpretation of soil micromorphological data. *Geoderma*, **8**, 81–94.

— and Walker, P. H. (1969). Weathering and soil development on a sequence of river terraces. *Aust. J. Soil Res.*, **20**, 293–305.

Broecker, W. S. (1965). Isotope geochemistry and the Pleistocene climatic record. In Wright, H. E. and Frey, D. G. (eds.), *The Quaternary of the United States.* Princeton University Press, 737–53.

Brown, E. H. (1970). Man shapes the earth. *Geogrl J.*, **136**, 74–85.

Browne, W. R. (1969). Geomorphology general notes. *J. geol. Soc. Aust.*, **16**(1), 559–72.

Brunner, H. (1970). Pleistozäne Klimaschwankungen im Bereich des östlichen Mysore-Plateaux (Südindien). *Geologie*, **19**, 72–82.

Brush, L. M. and Wolman, M. G. (1960). Knickpoint behaviour in noncohesive material: a laboratory study. *Bull. geol. Soc. Am.*, **71**, 59–74.

Büdel, J. (1957a). Die doppelten Einebnungsflächen in den feuchten Tropen. *Z. Geomorph. N. F.*, **1**, 201–88.

— (1957b). Die Flächenbildung in den feuchten Tropen und die Rolle fossiler solcher Flächer in anderen Klimazonen. *Verh. dt. Geogr. Tags.*, **31**, 89–109.

— (1969). Das System der klimagenetischen Geomorphologie. *Erdkunde*, **23**, 165–82.

Budyko, H. (1958). *The Heat Balance of the Earth's Surface.* Translated by Nina A. Stepanova. U.S. Dept of Commerce, Washington.

— (1965). Climate and waters, general survey. *Soviet Geogr.*, **6**(5–6), 298–303.

Butler, B. E. (1959). Periodic phenomena in landscapes as a basis for soil studies. *Soil Publs. CSIRO Aust.*, **14**.

— (1963). The place of soils in studies of Quaternary chronology in southern Australia. *Révue Géomorph. dyn.*, **14**, 160–70.

— (1967). Soil periodicity in relation to landform development in south-eastern Australia. In Jennings, J. N. and Mabbutt, J. A. (eds.), *Landform Studies from Australia and New Guinea.* A.N.U. Press, Canberra, 231–55.

Cailleux, A. (1948). Le ruissellement en pays tempéré non montagneux. *Annls. Géogr.*, **57**, 21–39.

— (1969). Ein Beitrag zu KRUMBEIN: Uber den Einfluss der Mikroflora auf die exogene Dynamik (Verwitterung und Krustenbildung). *Geol. Rdsch.*, **58**, 363–5.

— and Tricart, J. (1959). *Initiation à l'étude des sables et galets.* Tome 1. Centre de Documentation Universitaire, Paris.

Caine, N. (1968). The anlaysis of surface fabrics on talus by means of ground photography. *Arct. Alp. Res.*, **1**, 127–34.

Carlston, C. W. (1966). The effect of climate on drainage density and stream-flow. *Bull. Ass. int. Hydrol. Scient.*, **11**(3), 62–9.

Carson, M. A. (1969). Models of hillslope development under mass failure. *Geogrl Analysis*, **1**, 76–100.

— (1971). An application of the concept of threshold slopes to the Laramie Mountains, Wyoming. *Inst. Br. Geogr. spec. Publs.*, **3**, 31–48.

— and Kirkby, M. J. (1972). *Hillslope Form and Process*. Cambridge University Press.

— and Petley, D. J. (1970). The existence of threshold slopes in the denudation of the landscape. *Trans. Inst. Br. Geogr.*, **49**, 71–95.

Chang, H. Y., Simons, D. B. and Brooks, R. H. (1967). The effect of water detention structures on river and delta morphology. *Publs. Ass. int. Hydrol. Scient.*, **75**, 438–48.

Chapman, R. W. and Greenfield, M. A. (1949). Spheroidal weathering of igneous rocks. *Am. J. Sci.*, **247**, 407–29.

Charreau, G. (1969). Rain and erosion. *Sols. Afr.*, **1969**, 241–55.

Chitale, S. V. (1970). River channel patterns. *J. Hydraul. Div. Am. Soc. civ. Engrs.*, **96** (HY1), 201–21.

Cholley, A. (1950). Morphologie structurale et morphologie climatique. *Annls Géogr.*, **59**, 321–35.

Chorley, R. J. (1964). The nodal position and anomalous character of slope studies in geomorphological research. *Geogrl J.*, **130**, 70–3.

—, Beckinsale, R. F. and Dunn, A. J. (1973). *A History of the Study of Landforms or the Development of Geomorphology*. Vol. 2. *The Life and Work of William Morris Davis*. Methuen, London.

—, Dunn, A. J. and Beckinsale, R. P. (1964). *The History of the Study of Landforms*, Vol. 1. *Geomorphology before William Morris Davis*. Methuen, London.

—, and Kennedy, B. A. (1971). *Physical Geography — A Systems Approach*. Prentice-Hall International, London.

Clayton, K. M. (1971). Geomorphology — a study which spans the geology/geography interface. *J. geol. Soc.*, **127**, 471–6.

Cleaves, A. B. (1950). Sedimentation and highway engineering. In Trask, P. D. (ed.), *Applied Sedimentation*. Wiley, New York and Chapman and Hall, London, 127–46.

Colby, B. R. (1963). Fluvial sediments — a summary of source, transportation, deposition, and measurement of sediment discharge. *Bull. U.S. geol. Surv.*, **1181-A**.

— (1964). Scour and fill in sand-bed streams. *Prof. Pap. U.S. geol. Surv.*, **462-D**.

Coleman, J. M. (1968). Brahmaputra River: channel processes and sedimentation. *Sedim. Geol.*, **3**, 129–239.

Colinvaux, P. A. (1973). *Introduction to Ecology*. Wiley, New York.

Colville, J. S. and Holmes, J. W. (1972). Water table fluctuations under forest and pasture in a karstic region of southern Australia. *J. Hydrol.*, **17**, 61–80.

Cook, H. L. (1936). The nature and controlling variables of the water erosion process. *Proc. Soil Sci. Soc. Am.*, **1**, 487–94.

Cooke, R. U. and Warren, A. (1973). *Geomorphology in Deserts*. Batsford, London.

Cooray, P. G. (1967). *The Geology of Ceylon*. National Museums of Ceylon, Colombo.

Costin, A. B. (1971). Vegetation, soils, and climate in Late Quaternary Southeastern Australia. In Mulvaney, D. J. and Golson, J. (eds.), *Aboriginal Man and Environment in Australia*. A.N.U. Press, Canberra, 26–37.

Cotton, C. A. (1958). Fine-textured erosional relief in New Zealand. *Z. Geomorph., N. F.*, **2**, 187–210.

Crickmay, C. H. (1933). The later stages of the cycle of erosion. *Geol. Mag.*, **70**, 337–47.

— (1960). Lateral activity of a river of northwestern Canada. *J. Geol.*, **68**, 377–91.

— (1974). *The Work of the River.* Macmillan, London.

Crozier, M. J. (1968). Earthflows and related environmental factors of Eastern Otago. *J. Hydrol. (N.Z.)*, **7**(2), 4–12.

Cullen, R. M. and Donald, I. B. (1971). Residual strength determination in direct shear. *Proc. First Australia-New Zealand Conference on Geomechanics*, **1**, 1–10.

Culling, W.E.H. (1957). Multicyclic streams and the equilibrium theory of grade. *J. Geol.*, **65**, 259–74.

— (1963). Soil creep and the development of hillside slopes. *J. Geol.*, **71**, 127–61.

Dahlskog, S. (1966). Sedimentation and vegetation in a Lapland mountain delta. *Geogr. Annlr*, **48A**, 86–101.

Daley, B. (1972). Some problems concerning the early Tertiary climate of Southern Britain. *Palaeogeogr., Palaeoclimatol., Palaeoecol.*, **11**, 177–90.

Dalrymple, J. B., Blong, R. J. and Conacher, A. J. (1968). An hypothetical nine unit landsurface model. *Z. Geomorph., N. F.*, **12**, 60–76.

Dana, J. D. (1850). On denudation in the Pacific. *Am. J. Sci.* (Ser. 2.), **9**, 48–62. (reprinted in Schumm, S. A. (1972). *River Morphology.* Dowden Hutchinson, Ross, Stroudsburg, 24–39).

Daveau, S. (1965). Ruissellement et soutirages dans la haute vallée du Denkalé (Monts Loma, Sierra Leone). *Bull. Ass. Géogr. fr.*, 330–1, 20–7.

Davies, J. L. (1969). *Landforms of Cold Climates.* A.N.U. Press, Canberra.

Davis, I. H. (1965). Landform mapping in the Wingham district. *University of New England External Stud. Gaz.*, **9**(5), 2–6.

Davis, W. M. (1899). The geographical cycle. *Geogrl J.*, **14**, 481–504.

— (1905). Complications of the geographical cycle. *Report of the Eighth International Geogr. Congr. Washington 1904*, 150–63.

— (1923). The cycle of erosion and the summit level of the Alps. *J. Geol.*, **31**, 1–41.

de Geer, G. (1934). Equatorial Paleolithic varves in East Africa. *Geogr. Annlr*, **16**, 75–96.

Degens, E. T. (1965). *Geochemistry of Sediments.* Prentice-Hall, Englewood Cliffs.

Dejou, J. (1967a). Sur l'altération des granites à deux micas du massif de la Pierre-qui-Vire. *C. r. hebd. Séanc. Acad. Sci., Paris*, **264**, 37–40.

— (1967b). Présence de gibbsite et des gels aluminosiliciques dans la fraction argilleux extraite de quelques arènes et sols du massif de granite à 2 micas de la Pierre-qui-Vire. *C. r. hebd. Séanc. Acad. Sci., Paris*, **264**, 1973–6.

— (1967c). Le massif de granite à deux micas de la Pierre-qui-Vire. *Annls agron.*, **18**, 145–201.

—, Guyot, J., Pedro, G., Chaumont, C. and Antoine, H. (1968). Nouvelles données concernant la présence de gibbsite dans les formes d'altération superficielle des massifs granitiques (Cas du Cantal et du Limousin). *C. r. hebd. Séanc. Acad. Sci., Paris*, **266**, 1825–7.

—, —, —, — and — (1969). Nouvelles observations sur la présence de gibbsite dans les zones d'altération superficielle et les sols des massifs cristallins et cristallophylliens. *Annls agron.*, **20**, 639–43.

—, —, Morizet, J., Chaumont, C. and Antoine, H. (1970). Etude comparative de la composition minéralogique de la fraction argileux < 2μ extraite des sols et de zones d'altération sur granite a deux micas et micaschistes à biotite Limousin. *Sci. Sol.*, **2**, 3–14.

Demangeot, J. (1969). *Les milieux naturels tropicaux*. Centre de Documentation Universitaire, Paris.

de Martonne, E. (1940). Problèmes morphologiques du Brésil tropical atlantique. *Annls Géogr.*, **49**, 1–27 and 106–29.

— (1946). Géographie zonale. La zone tropicale. *Annls Géogr.*, **55**, 1–18.

— (1951). *Traité de Géographie Physique*, Tome 2, *Le Relief du Sol*. (9th ed.), Armand . Colin, Paris, 499–1052.

de Meis, R. M., and da Silva, J. X. (1968). Mouvements de masse récents à Rio de Janeiro: une étude de géomorphologie dynamique. *Révue Géomorph. dyn.*, **18**, 145–51.

Derbyshire, E. (1971). A synoptic approach to the atmospheric circulation of the last glacial maximum in southeastern Australia. *Palaeogeogr., Palaeoclimatol., Palaeoecol.*, **10**, 103–24.

— and Peterson, J. A. (1971). On the status and correlation of Pleistocene glacial episodes in south-eastern Australia. *Search*, **2**, 285–8.

Devdarini, A. S. (1967a). A plane mathematical model of the growth and erosion of an uplift. *Soviet Geogr.*, **8**, 183–98.

— (1967b). The profile of equilibrium and a regular regime. *Soviet Geogr.*, **8**, 168–83.

— and Greysukh, V. L. (1969). The role of cybernetic methods in the study and transformation of natural complexes. *Soviet Geogr.*, **10**, 14–23.

Doornkamp, J. C. and King, C.A.M. (1971). *Numerical Analysis in Geomorphology: an Introduction*. Arnold, London.

Douglas, I. (1967a). Erosion of granite terrains under tropical rain forest in Australia, Malaysia and Singapore. *Publs. Ass. int. Hydrol. Scient.*, **75**, 31–9.

— (1967b). Man, vegetation and the sediment yields of rivers. *Nature, Lond.*, **215**, 925–8.

— (1968a). Erosion in the Sungei Gombak catchment, Selangor, Malaysia. *J. trop. Geogr.*, **26**, 1–16.

— (1968b). The effects of precipitation chemistry and catchment area lithology on the quality of river water in selected catchments in eastern Australia. *Earth Sci. J.*, **2**, 126–44.

— (1969). The efficiency of humid tropical denudation systems. *Trans. Inst. Brit. Geogrs.*, **46**, 1–16.

— (1975). The impact of urbanization on river systems. *Proc. I.G.U. Regional Conf. 8th New Zealand Geog. Conf.*, 307–17.

Dury, G. H. (1959). *The Face of the Earth.* Penguin Books, Harmondsworth.

— (1965). Theoretical implications of underfit streams. *Prof. Pap. U.S. geol. Surv.*, **452-G**.

— (1966a). The concept of grade. In Dury, G. H. (ed.), *Essays in Geomorphology.* Heinemann, London, 211–33.

— (1966b). Incised valley meanders of the lower Colo River, New South Wales. *Aust. Geogr.*, **10**, 17–25.

— (1967). Some channel characteristics of the Hawkesbury River, New South Wales. *Aust. geogrl Stud.*, **5**, 135–49.

— (1969a). Rational descriptive classification of duricrusts. *Earth Sci. J.*, **3**, 77–86.

— (1969b). Hydraulic geometry. In Chorley, R. J. (ed.), *Water, Earth and Man.* Methuen, London, 319–29.

— (1970a). A re-survey of part of the Hawkesbury River, New South Wales, after one hundred years. *Aust. geogrl Stud.*, **8**, 121–32.

— (1970b). General theory of meandering valleys and underfit streams. In Dury, G. H. (ed.), *Rivers and River Terraces.* Macmillan. London, 264–75.

Dutton, C. E. (1880). *Report on the Geology of the High Plateaus of Utah.* U.S. Geographical and Geological Survey of the Rocky Mountain Region, Washington.

Eden, M. J. (1964). The savanna ecosystem — northern Rupununi, British Guiana, *McGill Univ. Savanna Research Project, Savanna Research Ser.*, 1.

Edmonds, E. A., Poole, E. G. and Wilson, V. (1965). Geology of the country around Banbury and Edge Hill. *Mem. geol. Surv. U.K.*, **201**.

Eidmann, F. E. (1962). Über den Wasserhaushault von Buchen- und Fichten-beständen. *IUFRO — Congress, Vienna 1961*, 2 Teil, Bd 1, 11–14.

Einstein, H. A. (1964). River sedimentation. In Chow, Ven Te (ed.), *Handbook of Applied Hydrology.* McGraw-Hill, New York, 17–35 to 17–67.

Emiliani, C. (1971). The amplitude of Pleistocene climatic cycles at low latitude and the isotopic composition of glacial ice. In Turekian, K. K. (ed.), *Late Cenozoic Glacial Ages.* Yale University Press, New Haven, 183–97.

— and Shackleton, N. J. (1973). The Brunhes Epoch: isotopic paleotemperatures and geochronology. *Science*, **183**, 511–14.

Engels, H. (1905). Untersuchungen über die Bettausbildung..., *Z. Bauw.*, **55**.

Emmett, W. W. and Hadley, R. F. (1968). The vigil network: preservation and access of data. *Circ. U.S. geol. Surv.*, **460-C**.

Evans, I. S. (1970). Salt crystallization and rock weathering: a review. *Révue Géomorph. dyn.*, **19**, 153–77.

Eyles, R. J. (1967). Laterite in Kerdau, Pahang, Malaya. *J. trop. Geogr.*, **25**, 18–23.

Fink, J. (1961). Der östliche Teil des nördlichen Alpenvorlandes. *Mitt. österreichen bodenkundlichen Gesell.*, **6**, 26–51.

— (1965). The Pleistocene in eastern Austria. *Spec. Pap. geol. Soc. Am.*, **84**, 179–99.

Fisk, H. N. (1944). *Geological Investigation of the Alluvial Valley of the Lower Mississippi*. Mississippi River Comm., Waterways Experiment Station, Vicksburg.

— (1947). *Fine-grained Alluvial Deposits and their Effects on Mississippi River Activity*. Mississippi River Comm., Waterways Experiment Station, Vicksburg.

Flenley, J. R. (1974). In search of the past. *Geogrl Mag.*, **47**, 162–9.

Fulton, R. J. and Pullen, M.J.L.T. (1969). Sedimentation in Upper Arrow Lake, British Columbia. *Can. J. Earth Sci.*, **6**, 785–91.

Galloway, R. W. (1963). Geomorphology of the Hunter Valley. *Land Res. Ser. CSIRO Aust.*, **8**, 90–102.

— (1965). A note on world precipitation during the last glaciation. *Eiszeitalter und Gegenwart.*, **16**, 76–7.

— (1971). Evidence for Late Quaternary climates. In Mulvaney, D. J. and Golson, J. (eds.), *Aboriginal Man and Environment in Australia*. A.N.U. Press, Canberra, 14–25.

Garner, H. F. (1966). Derangement of the Rio Caroni, Venezuela. *Révue Géomorph. dyn.*, **16**, 54–83.

— (1968). Climatic geomorphology. In Fairbridge, R. W. (ed.), *Encyclopaedia of Geomorphology*. Reinhold, New York, 129–31.

— (1974). *The Origin of Landscapes: A Synthesis of Geomorphology*. Oxford University Press, New York.

Garrels, R. M. and Mackenzie, F. J. (1971). *Evolution of the Sedimentary Rocks*. Norton, New York.

Geddes, A. (1960). The alluvial morphology of the Indo-Gangetic Plain: its mapping and geographical significance. *Trans. Inst. Br. Geogr.*, **28**, 253–76.

Geyl, W. F. (1968). Tidal stream action and sea level change as one cause of valley meanders and underfit streams. *Aust. geogrl Studies*, **6**, 24–42.

Gibbs, R. J. (1967). The geochemistry of the Amazon River Basin: Part I: The factors that control the salinity and the composition and concentration of suspended solids. *Bull. geol. Soc. Am.*, **78**, 1203–32.

Gilbert, G. K. (1877). *Report on the Geology of the Henry Mountains*. U.S. Department of the Interior, Washington.

— (1880). *Report on the Geology of the Henry Mountains* (2nd ed.). U.S. Department of the Interior, Washington.

Godard, A. (1966). Les 'tors' et le problème de leur origine. *Révue géogr. de l'Est*, **6**, 153–70.

— (1972). Quelques enseignements apportés par le massif central français dans l'étude géomorphologique des socles cristallins. *Révue Géogr. phys. Géol. dyn.*, **14**, 265–96.

Goede, A. (1965). Geomorphology of the Buckland Basin, Tasmania. *Pap. Proc. R. Soc. Tasmania*, **99**, 133–54.

— (1972). Discontinuous gullying of the Tea Tree Rivulet, Buckland, eastern Tasmania. *Pap. Proc. R. Soc. Tasmania*, 106, 5–14.

Gossman, H. (1970). Theorien zur Hangentwicklung in verschiedenen Klimazonen. *Würzburger Geogr. Arb.*; 31.

Graf, W. H. (1971). *Hydraulics of Sediment Transport*. McGraw-Hill, New York.

Grant, P. J. (1965). Major regime changes of the Tukituki River, Hawke's Bay since about 1650. *J. Hydrol. (N.Z.)*, 4(1), 17–30.

Gregory, K. J. (1971). Drainage density changes in south west England. In Gregory. K. J. and Ravenhill, W.L.D. (eds.), *Exeter Essays in Geography*. The University, Exeter, 33–53.

— and Walling, D. E. (1968). The variation of drainage density within a catchment. *Bull. Ass. int. Hydrol. Scient.*, 13, 61–8.

— and — (1971). Field measurements in the drainage basin. *Geography*, 56, 277–92.

Gupta, A. (1975). Stream characteristics in eastern Jamaica, an environment of seasonal flow and large floods. *Am. J. Sci.*, 275, 825–47.

Haantjens, H. A. (1970). Soils of the Goroka-Mount Hagen area. *Land Res. Ser. CSIRO Aust.*, 27, 80–103.

Hack, J. T. (1957). Studies of longitudinal stream profiles in Virginia and Maryland. *Prof. Pap. U.S. geol. Surv.*, 294B.

— (1960). Interpretation of erosional topography in humid temperate regions. *Am. J. Sci.*, 258-A, 80–97.

— (1965). Geomorphology of the Shenandoah Valley Virginia and West Virginia and origin of the residual ore deposits. *Prof. Pap. U.S. geol. Surv.*, 484.

— and Goodlett, J. C. (1960). Geomorphology and forest ecology of a mountain region in the central Appalachians. *Prof. Pap. U.S. geol. Surv.*, 347.

Haggett, P., Chorley, R. J. and Stoddart, D. R. (1965). Scale standards in geographical research. A new measure of areal magnitude. *Nature, Lond.*, 205, 844–7.

Hamvas, F. (1967). A balatoni parterózió vizgálta. *Hidrologia Közlöny*, 47, 560–3.

Harbaugh, J. W. and Bonham-Carter, G. (1970). *Computer Simulation in Geology*. Wiley-Interscience, New York.

Harding, S. T. (1942). Lakes. In Meinzer, O. E. (ed.), *Hydrology*. Dover Publications, New York, 220–43.

Harvey, A. M. (1969). Channel capacity and the adjustment of streams to hydrologic regime. *J. Hydrol.*, 8, 82–98.

— (1975). Some aspects of the relations between channel characteristics and riffle spacing in meandering streams. *Am. J. Sci.*, 275, 470–8.

Hastenrath, S. (1971). On snow line depression and atmospheric circulation in the tropical Americas during the Pleistocene. *S. Afr. geogrl J.*, 53, 53–69.

Hayford, J. F. (1922). Effects of winds and barometric pressures on the Great Lakes. *Carnegie Inst. Wash. Publ.*, 317.

Hem, J. D. (1959). Study and interpretation of the chemical characteristics of natural water. *Wat. -Supply Pap., Wash.*, 1473.

Henderson, F. M. (1963). Stability of alluvial channels. *Trans. Amer. Soc. civ. Engrs.*, **128**, 657–720.

Hettner, A. (1928). *Die Oberflächenformen des Festlandes.* Trubner, Stuttgart.

— (1972). *The Surface Features of the Land: Problems and Methods of Geomorphology.* Macmillan, London.

Hewlett, J. D. and Hibbert, A. R. (1963). Moisture and energy conditions within a sloping soil mass during drainage. *J. geophys. Res.*, **68**, 1081–7.

— and Nutter, W. L. (1970). The varying source area of streamflow from upland basins. *Proc. Symp. Interdisciplinary Aspects Watershed Management, Montana,* Amer. Soc. Civ. Engnrs., 65–8.

Heydenreich, L. H. (1954). *Leonardo da Vinci.* Allen and Unwin, London.

Hickin, E. J. (1969). A newly-identified process of point bar formation in natural streams. *Am. J. Sci.*, **267**, 999–1010.

— (1970). The terraces of the Lower Colo and Hawkesbury drainage basins, New South Wales. *Aust. Geogr.*, **11**, 278–87.

Hills, E. S. (1940). *The Physiography of Victoria.* Whitcombe and Tombs, Melbourne.

Hinz, W. (1970). *Silikate: Grundlagen der Silikatwissenschaft und Silikattechnik Band 1.* Verlag für Bauwesen, Berlin.

Hjulström, F. (1935). Studies of the morphological activity of rivers as illustrated by the River Fyris. *Bull. geol. Inst. Univ. Uppsala*, **25**, 221–528.

Ho, R. (1962). I.G.U. Guide to Tours Malayan Regional Conference. Geography Dept, University of Malaya (unpublished).

Hopkins, B. (1965). Vegetation of the Olokemeji Forest Reserve, Nigeria, III, The microclimates with special reference to their seasonal changes. *J. Ecol.*, **53**, 125–38.

Horton, R. E. (1927). *Report of Engineering Board of Review of Sanitary District of Chicago on Lake-lowering Controversy and a Program of Remedial Measures, Part 3, Appendix 2, Hydrology of the Great Lakes.*

— (1932). Drainage basin characteristics. *Trans. Am. geophys. Un.*, **21**, 522–41.

— (1945). Erosional development of streams and their drainage basins: Hydrophysical approach to quantitative morphology. *Bull. geol. Soc. Am.*, **56**, 275–370.

Imeson, A. C. (1970). Variations in sediment production from three East Yorkshire catchments. In Taylor, J. A. (ed.), *The Role of Water in Agriculture.* Pergamon, Oxford, 39–56.

Jackson, W. D. (1965). Vegetation. In Davies, J. L. (ed.), *Atlas of Tasmania.* Lands and Survey Department, Hobart, 30–5.

Jahn, A. (1956). Slope research studies in Poland. *Przeglad Geogr.*, **28**, 94–100.

Jéjé, L. K. (1972). Landform development at the boundary of sedimentary and crystalline rocks in southwestern Nigeria. *J. trop. Geogr.*, **34**, 25–33.

Jennings, J. N. (1963). Floodplain lakes in the Ka Valley, Australian New Guinea. *Geogrl. J.*, **129**, 187–90.

— (1965). Man as a geological agent. *Aust. J. Sci.*, **28**, 150–6.

— (1971). *Karst*. A.N.U. Press, Canberra.

Joshi, R. (1972). The characteristics of the Pleistocene climatic events in the Indian sub-continent. A land of monsoon climate. In Ters, M. (ed.), *Études sur le Quaternaire dans le Monde*, Vol. 1.

Kajetanowicz, Z. (1958). La dépendance de la grosseur de la granulation du matériel du lit dans les rivières montagneuses de leurs qualités physiographiques. *Publs. Ass. int. Hydrol. scient.*, **43**, 323–6.

Kaye, C. A. (1950). Principles of soil mechanics as viewed by a geologist. In Trask, P. D. (ed.), *Applied Sedimentation*. Wiley, New York and Chapman and Hall, London, 93–112.

Keller, E. A. (1971). Pools, riffles, and meanders: discussion. *Bull. geol. Soc. Am.*, **82**, 279–80.

Keller, W. D. (1954). The energy factor in sedimentation. *J. sedim. Petrol.*, **24**, 62–8.

Kelly, P. W. (1972). Are the pools and riffles of the stream leading into Oxford Falls regularly spaced? Unpublished Geography IIIB Research Methods Exercise, University of New England.

Kenney, T. C. (1967). The influence of mineral composition on the residual strength of natural soils. *Proc. Geotech. Conf., Oslo, 1967*, **1**, 123–30.

Kershaw, A. P. (1975). Late Quaternary vegetation and climate in northeastern Australia. *Bull. Roy. Soc. N.Z.*, **13**, 181–7.

King, C.A.M. (1966). *Techniques in Geomorphology*. Arnold, London.

King, L. C. (1950). The study of the world's plainlands: a new approach in geomorphology. *Q. Jl geol. Soc. Lond.*, **106**, 101–31.

— (1953). Canons of landscape evolution. *Bull. geol. Soc. Am.*, **64**, 721–53.

Kirkby, M. J. (1967). Measurement and theory of soil creep. *J. Geol.*, **7**, 359–78.

— (1969). Infiltration, throughflow and overland flow. In Chorley, R. J. (ed.), *Water, Earth and Man*. Methuen, London, 215–28.

— (1971). Hillslope process-response models based on the continuity equation. *Inst. Br. Geogr. spec. Publs.*, **3**, 15–30.

— and Chorley, R. J. (1967). Throughflow, overland flow and erosion. *Bull. Ass. int. Hydrol. scient.*, **12**, 5–21.

Knighton, A. D. (1975). Variations in at-a-station hydraulic geometry. *Am. J. Sci.*, **275**, 186–218.

Knox, J. C. (1972). Valley alluviation in southwestern Wisconsin. *Ann. Ass. Am. Geogr.*, **62**, 401–10.

Kolosius, H. J. (1971). Rigid boundary hydraulics for steady flow. In Shen, H. W. (ed.), *River Mechanics I*. The Editor, Fort Collins, 3–1 to 3–51.

Koziejowa, U. (1963). Denudjacja stokow w rooznym cyklu klimatyeznym. *Acta Geogr. Lodz*, 16, 7–56 (French summary: Érosion des versants au cours du cycle climatique d'une année).

Krumbein, W. C. and Graybill, F. A. (1965). *An Introduction to Statistical Models in Geology*. McGraw-Hill, New York.

Krumbein, W. E. (1969). Über den Einfluss der Mikroflora auf die exogene Dynamik (Verwitterung und Krustenbildung). *Geol. Rdsch.*, 58, 333–63.

— and Graybill, F. A. (1965). *An Introduction to Statistical Models in Geology*. McGraw-Hill, New York.

Lamb, H. H. (1966). Climate in the 1960's: changes in the world's wind circulation reflected in prevailing temperatures, rainfall patterns and the levels of the African lakes. *Geogrl J.*, 132, 183–212.

Lamotte, M. and Rougerie, G. (1962). Les apports allochtones dans la genèse des cuirasses ferrugineuses. *Révue Géomorph. dyn.*, 13, 145–60.

Lane, E. W. and Carlson, E. J. (1954). Some observations on the effect of particle shape on the movement of coarse sediment. *Trans. Am. geophys. Un.*, 35, 453–62.

Langbein, W. B. and Leopold, L. B. (1966). River meanders — theory of minimum variance. *Prof. Pap. U.S. geol. Surv.*, 422 H.

Leopold, L. B. (1969). The rapids and pools — Grand Canyon. *Prof. Pap. U.S. geol. Surv.*, 669, 131–45.

—, Emmett, W. W. and Myrick, R. W. (1966). Channel and hillslope processes in a semi-arid area, New Mexico. *Prof. Pap. U.S. geol. Surv.*, 352 G, 193–253.

— and Langbein, W. B. (1962). The concept of entropy in landscape evolution. *Prof. Pap. U.S. geol. Surv.*, 500-A.

— and Maddock, T. (1953). The hydraulic geometry of stream channels and some physiographic implications. *Prof. Pap. U.S. geol. Surv.*, 252.

— and Wolman, M. G. (1957). River channel patterns — braided, meandering, and straight. *Prof. Pap. U.S. geol. Surv.*, 282 B, 39–85.

— and — (1960). River meanders. *Bull. geol. Soc. Am.*, 71, 769–94.

—, — and Miller, J. P. (1964). *Fluvial Processes in Geomorphology*. Freeman, San Francisco.

Lewis, A. J. (1974). Geomorphic-geologic mapping from remote sensors. In Estes, J. E. and Senger, L. W. (eds.), *Remote Sensing: Techniques for Environmental Analysis*. Hamilton, Santa Barbara, 105–26.

Lewis, L. A. (1966). The adjustments of some hydraulic variables at discharges less than one cfs. *Prof. Geogr.*, 18, 230–4.

Likens, G. E. and Bormann, F. H. (1974). Linkages between terrestrial and aquatic ecosystems, *Bioscience*, 24, 447–56.

Lineham, W. (1951). Traces of a bronze age culture associated with iron age implements in the regions of Klang and the Tembeling, Malaya. *J. Malay. Brch R. Asiat. Soc.*, 24(3), 1–59.

Linton, D. L. (1955). The problem of tors. *Geogrl J.*, 121, 470–87.

Looman, H. (1956). Observations about some differential equations concerning recession of mountain slopes. *Proc. K. ned. Akad. Wet., Ser. B.*, **59**, 259–84.

Louis, H. (1964). Über Rumpfflächen und Talbildung in den Wechselfeuchten Tropen besonders nach Studien in Tanganyika. *Z. Geomorph. N. F.*, **8**, Sonderheft, 43–70.

— (1968). *Allgemeine Geomorphologie.* (2e. Auflage), De Gruyter, Berlin.

Lucas, J. (1962). La transformation des minéraux origineux dans la sédimentation, études sur les argiles du Trias. *Mém. Serv. Carte géol. Als.-Lorr.*, **23**.

Lustig, L. K. (1965). Clastic sedimentation in Deep Springs Valley California. *Prof. Pap. U.S. geol. Surv.*, **352-F**, 131–92.

— and Busch, R. D. (1967). Sediment transport in Cache Creek drainage basin in the Coast Ranges west of Sacramento, California. *Prof. Pap. U.S. geol. Surv.*, **562-A**.

Mabbutt, J. A. (1977). *Desert Landforms.* A.N.U. Press, Canberra.

— and Scott, R. M. (1966). Periodicity of morphogenetic and soil formation in a savannah landscape near Port Moresby, Papua. *Z. Geomorph., N. F.*, **10**, 69–89.

Mackin, J. H. (1948). Concept of the graded river. *Bull. geol. Soc. Amer.*, **59**, 463–512.

Mainguet, M. (1972). *Le modèle des grès: problèmes généraux.* Institut Géographique National, Paris.

Markov, K. K. (1969). Geographical regions and zones and their Quaternary development. In Wright, J. E. Jr (ed.), *Quaternary Geology and Climate* (Proc. VII Congress INQUA, Vol. 16, Publs, natn. Acad. Sci., 1701), 3–5.

Melton, M. A. (1960). Intravalley variation in slope angles related to microclimate and erosional environment. *Bull. geol. Soc. Am.*, **71**, 133–44.

Mescheryakov, Y. A. (1959). Primeie geomorfologicheskix methadov pri poiskax gaza i nefti. *Otd geol-geogr. nauk A. M. S.S.S.R. materialy 2-go geomorfol. soveshchaniya, Moscow.*

Meynier, A. (1961). Glissements de terrain dans le bassin permien de Brive. *Révue Géomorph. dyn.*, **12**, 130–6.

Michel, P. (1973). Les bassins des fleuves Sénégal et Gambie: étude géomorphologique. *Mémoires ORSTOM*, **63**.

— and Assémien, P. (1970). Études sédimentologiques et palynologiques des sondages de Bogué (Basse Vallée) du Sénégal et leur interprétation morphoclimatique. *Révue Géomorph. dyn.*, **19**, 97–113.

Miller, A. A. (1964). *The Skin of the Earth.* Methuen University Paperbacks, London.

Miller, R. L. and Kahn, J. S. (1962). *Statistical Analysis in the Geological Sciences.* Wiley, New York.

Millot, G. (1965). *Géologie des argiles.* Masson, Paris.

—, Cogné, J., Jeanette, D., Besnus, Y., Monnet, B., Guri, F., and Schimpf, A. (1967). La maladie des grès de la cathédrale de Strasbourg. *Bull. Serv. Carte géol. Als.-Lorr.*, **20**, 131–57.

Monkhouse, F. J. and Wilkinson, H. R. (1970). *Maps and Diagrams.* (3rd ed.), Methuen, London.

Moore, P. D. and Bellamy, D. J. (1974). *Peatlands.* Elek, London.

Morgan, M. A. (1969). Overland flow and man. In Chorley, R. J. (ed.), *Water, Earth and Man.* Methuen, London, 239–55.

Morgan, R.P.C. (1972). Observations of factors affecting the behaviour of a first-order stream. *Trans Inst. Br. Geogrs.*, **56**, 171–85.

Mörth, H. (1965). Investigations into the meteorological aspects of the variations in the level of Lake Victoria. Unpublished report of the East African Meteorological Department, Nairobi.

Moss, A. J. (1972). Bed-load sediments. *Sedimentology*, **18**, 159–220.

Moushinho de Meis, M. R. (1971). Upper Quaternary process change of the middle Amazon area. *Bull. geol. Soc. Am.*, **82**, 1073–8.

Mulvaney, D. J. (1969). *The Prehistory of Australia.* Thames and Hudson, London.

Negev, M. (1969). Analysis of data on suspended sediment discharge in several streams in Israel. *State of Israel hydrological Serv. Hydrol. Pap.*, **12**.

Nossin, J. J. (1964). Geomorphology of the surroundings of Kuantan (Eastern Malaya). *Geol. Mijnbouw.*, **44**, 157–82.

Odum, E. P. (1963). *Ecology.* Holt, Rinehart and Winston, New York.

Ofomata, G.E.K. (1966). Quelques observations sur l'éboulement d'Awgu, Nigeria oriental. *Bull. Inst. franç., Afric. Noire, Ser. A.*, **28**, 433–43.

Ollier, C. D. (1963). Insolation weathering: examples from Central Australia. *Am. J. Sci.*, **261**, 376–81.

— (1965). Some features of granite weathering in Australia. *Z. Geomorph. N. F.*, **9**, 285–3–4.

— (1969a). *Weathering.* Oliver and Boyd, Edinburgh.

— (1969b). *Volcanoes.* A.N.U. Press, Canberra.

Ovington, J. D. (1962). Quantitative ecology and the woodland ecosystem concept. *Adv. ecol. Res.*, **1**, 103–92.

— (1964). Nutrient cycling in woodlands. In Halsworth, E. G. and Crawford, D. V. (eds.), *Experimental Pedology.* Butterworths, London, 208–18.

Owen, G. (1951). A provisional classification of Malayan soils. *J. Soil Sci.*, **2**, 20–42.

Pain, C. F. (1969). The effect of some environmental factors on rapid mass movement in the Hunua Ranges, New Zealand. *Earth Sci. J.*, **3**, 101–7.

— and Bowler, J. M. (1973). Denudation following the November 1970 earthquake at Madang, Papua New Guinea. *Z. Geomorph. N. F., Supplbd* **18**, 92–104.

— and Hosking, P. L. (1970). The movement of sediment in a channel in relation to magnitude and frequency concept — a New Zealand example. *Earth Sci. J.*, **4**, 17–23.

Pardé, M. (1954). Sur les érosions superficielles, les transports solides et les remblaiements effectués par les eaux courantes. *Publs. Ass. int. Hydrol. scient.*, 36, 194–206.

— (1958). Transports énormes de matériaux de fond par certaines rivières. *Publs. Ass. int. Hydrol. scient.*, 43, 360–70.

Paton, T. R. and Williams, M.A.J. (1972). The concept of laterite. *Ann. Ass. Am. Geogr.*, 62, 42–56.

Peel, R. F. (1941). The North Tyne valley. *Geogrl J.*, 98, 5–19.

Pels, S. (1971). Radio-carbon datings of ancestral river sediments on the Riverine plain of south-eastern Australia and their interpretation. *J. Proc. R. Soc. N.S.W.*, 102, 189–95.

Penck, W. (1925). Die Piedmontflächen des südlichen Schwarzwaldes. *Z. Gesell. Erdkunde Berlin*, 60, 81–108.

— (1953). *Morphological Analysis of Landforms*. Translated by H. Czech and K. C. Boswell. Macmillan, London.

Pereira, H. C. (1973). *Land Use and Water Resources in Temperate and Tropical Climates*. Cambridge University Press.

Philbrick, S. S. (1970). Horizontal configuration and the rate of erosion of Niagara Falls. *Bull. geol. Soc. Am.*, 81, 3723–32.

Phipps, R. L. (1974). The soil creep-curved tree fallacy. *J. Res. U.S. geol. Surv.*, 2, 371–7.

Pitty, A. F. (1965). A study of some escarpment gaps in the southern Pennines. *Trans. Inst. Br. Geogr.*, 37, 127–45.

— (1966). Some problems in the location and delimitation of slope-profiles. *Z. Geomorph. N. F.*, 10, 454–61.

— (1969). A scheme for hillslope analysis I, Initial considerations and calculations. *Univ. Hull Occ. Pap. Geogr.*, 9.

— (1970). A scheme for hillslope analysis II, Indices and tests for differences. *Univ. Hull Occ. Pap. Geogr.*, 17.

— (1971). *Introduction to Geomorphology*. Methuen, London.

Pop, Gh. (1964). Importance of the periodically wet tropical palaeoclimates in the genesis of some levelled surfaces in the Apuseni mountains. *Rev. Roumaine Géol. Géogr., Ser. Géogr.*, 8, 159–65.

Popov, I. V. (1964). Hydrological principles of the theory of channel processes and their use in hydrological planning. *Soviet Hydrol.*, 1964, 188–95.

Porrenga, D. H. (1967). Clay mineralogy and geochemistry of recent marine sediments in tropical areas. *Publs. Fysisch-Geogr. Lab. Univ. Amsterdam*, 9.

Prior, D. B., Stephens, N. and Douglas, G. R. (1971). Some examples of mudflow and rockfall activity in north-east Ireland. *Inst. Br. Geogr. Spec. Publs.*, 3, 129–40.

Pullar, W. A. (1965). Chronology of flood plains, fans and terraces in the Gisborne and Bay of Plenty districts. *Proc. 4th N.Z. Geog. Conf.*, 77–81.

— (1967). Uses of volcanic ash beds in geomorphology. *Earth Sci. J.*, 1, 164–77.

—, Pain, C. F. and Johns, R. J. (1967). Chronology of terraces, floodplains, fans and dunes in the Whakatone Valley. *Proc. 5th N.Z. Geog. Conf.*, 175–80.

Rahn, P. H. (1969). The relationship between natural forested slopes and angles of repose for sand and gravel. *Bull. geol. Soc. Am.*, 80, 2123–8.

Ralph, E. K. and Michael, H. N. (1969). University of Pennsylvania radiocarbon dates XII. *Radiocarbon*, 11, 469–81.

Raynal, R. and Nonn, H. (1968). Glacis étagés et formations quaternaires de Galice orientale et de Leon: quelques observations et données nouvelles. *Rêvue Géomorph. dyn.*, 18, 97–117.

Reiner, E. and Mabbutt, J. A. (1968). Geomorphology of the Wewak-Lower Sepik Area. *Land Res. Ser. CSIRO Aust.*, 22, 61–71.

Richards, K. S. (1973). Hydraulic geometry and channel roughness — a nonlinear system. *Am. J. Sci.*, 273, 877–96.

Robinson, D. A. (1971). Aspects of the geomorphology of the Central Weald. In Williams, R.B.G. (ed.), *Guide to Sussex Excursions*. Inst. Brit. Geogrs. Conf., Univ. Sussex, Brighton, 51–60.

Rochefort, M. and Tricart, J. (1959). Rôle de l'écoulement subsuperficiel dans l'alimentation de certains cours d'eau. *C. r. hebd. Séanc. Acad. Sci., Paris*, 248, 267–70.

Roe, F. W. (1953). The geology and mineral resources of the neighbourhood of Kuala Selangor and Rasa, Selangor, Federation of Malaya, with an account of the Batu Arang coal-field. *Geol. Surv. Dept Fed. Malaya Mem.*, 7.

Ronai, A. (1965). Neotectonic subsidences in the Hungarian basin. *Geol. Soc. Am. Spec. Pap.*, 84, 219–32.

Rougerie, G. (1960). Le façonnement actuel des modelés en Côte d'Ivoire forestière. *Mem. Inst. fr. Afric. Noire*, 58.

— (1965). Mobilisation des débris et pertes de substance dans la région du Ballon d'Alsace. *Rêvue géogr. de l'Est*, 5, 483–97.

Russell, R. J. (1954). Alluvial morphology of Anatolian rivers. *Ann. Ass. Am. Geogr.*, 44, 363–91.

Ruxton, B. P. (1967). Slopewash under mature primary rainforest in northern Papua. In Jennings, J. N. and Mabbutt, J. A. (eds.), *Landform. Studies from Australia and New Guinea*. A.N.U. Press, Canberra, 85–94.

— (1969). Geomorphology of the Kerema-Vailala area. *Land Res. Ser. CSIRO Aust.*, 23, 65–76.

— and Berry, L. (1957). Weathering of granite and associated erosional features in Hong Kong. *Bull. geol. Soc. Am.*, 68, 1263–92.

— and — (1961). Weathering profiles and geomorphic position on granite in two tropical regions. *Rêvue Géomorph. dyn.*, 12, 16–31.

Santos-Cayade, J. and Simons, D. B. (1973). River Response. In Shen, H. W. (ed.), *Environmental Impact on Rivers (River Mechanics III)*. The Editor, Fort Collins, 1–1 to 1–25.

Sapper, K. (1935). Geomorphologie der feuchten Tropen. *Geogr. Schr.*, 7.

Savigear, R.A.G. (1965). A technique of morphological mapping. *Ann. Ass. Am. Geogr.*, 55, 514–38.

— (1967). The analysis and classification of slope profile forms. In Maear, P. (ed.), *L'évolution des versants*. Ve Rapport de la Commission pour l'Étude des versants de l'Union Géographique Internationale, Liège, 271–90.

Scheidegger, A. E. (1961). *Theoretical Geomorphology*. Springer-Verlag, Berlin.

— (1965). The algebra of stream order numbers. *Prof. Pap. U.S. geol. Surv.*, 525-B, B 187–B189.

— (1970). *Theoretical Geomorphology*. (2nd ed.), Allen and Unwin, London and Springer-Verlag, Berlin.

— and Langbein, W. B. (1966). Probability concepts in geomorphology. *Prof. Pap. U.S. geol. Surv.*, 500-C.

Schumm, S. A. (1956a). Evolution of drainage systems and slopes in badlands at Perth Amboy, New Jersey. *Bull. geol. Soc. Am.*, 67, 597–646.

— (1956b). Role of creep and rainwash in the retreat of badland slopes. *Am. J. Sci.*, 254, 693–706.

— (1963a). A tentative classification of river channels. *Circ. U.S. geol. Surv.*, 477.

— (1963b). The sinuosity of alluvial rivers on the Great Plains. *Bull. geol. Soc. Am.*, 74, 1089–100.

— (1963c). The disparity between present rates of denudation and orogeny. *Prof. Pap. U.S. geol. Surv.*, 454 H.

— (1967). Meander wavelength of alluvial rivers. *Science*, 157, 1549–50.

— (1968). River adjustment to altered hydrologic regimen — Murrumbidgee River and palaeochannels, Australia. *Prof. Pap. U.S. geol. Surv.*, 598.

— (1971). Fluvial geomorphology: the historical perspective. In Shen, H. W. (ed.), *River Mechanics I*. The Editor, Fort Collins 4–1 to 4–30.

— and Lichty, R. W. (1965). Time, space and causality in geomorphology. *Am. J. Sci.*, 263, 110–19.

Schwab, G. O., Frevert, R. K., Barnes, K. K. and Edminster, T. W. (1971). *Elementary Soil and Water Engineering*. Wiley, New York.

Schwarzbach, M. (1967). Isländische Wasserfälle und eine genetische Systematik des Wasserfälles überhaupt. *Z. Geomorph., N. F.*, 11, 377–417.

Searle, A. B. and Grimshaw, R. W. (1959). *The Chemistry and Physics of Clays and Other Ceramic Materials*. (3rd ed.), Benn, London.

Selby, M. J. (1966). Some slumps and boulder fields near Whitehall. *J. Hydrol. (N.Z.)*, 5(2), 35–44.

— (1967a). *The Surface of the Earth*. Volume I. Cassell, London.

— (1967b). Aspects of the geomorphology of the greywacke ranges of the lower and middle Waikato basins. *Earth Sci. J.*, 1, 37–58.

— (1967c). Erosion by high intensity rainfalls in the lower Waikato. *Earth Sci. J.*, 1, 153–6.

— (1968). Morphometry of drainage areas of pumice lithology. *Proc. 5th N.Z. Geogr. Conf.*, 169–74.

Selivancy, G. D. and Svarichevskaya, Z. A. (1967). Vopopad na r. Sablinke (Leningradskaya oblast). *Leningradskiy Univ. Vestnik: Geol. Geogr.*, 4, 152–8.

Sellin, R.H.J. (1969). *Flow in Channels*. Macmillan, London.

Shelton, J. S. (1966). *Geology Illustrated*. Freeman, San Francisco.

Shen, H. W. (1971). Stability of alluvial channels. In Shen, H. W. (ed.), *River Mechanics III*. The Editor, Fort Collins, 16–1 to 16–33.

Sherlock, R. L. (1922). *Man as a Geological Agent*. Witherby, London.

Sherman, G. D. (1952). The genesis and morphology of the alumina-rich laterite clays. *Amer. Inst. Min. Metal Eng., Problems of Clay and Laterite Genesis*, 154–61.

Shoobert, J. (1968). Australian landform examples No. 12. Underfit streams of the Osage type: Head of the Port Hacking River. *Aust. Geogr.*, 10, 523–4.

Shreve, R. L. (1967). Infinite topologically random channel networks. *J. Geol.*, 75, 178–86.

Siever, R. (1959). Petrology and geochemistry of silica cementation in some Pennsylvanian sandstones. In Ireland, H. A. (ed.), Silica in sediments: a symposium. *Soc. Econ. Paleontologists and Mineralogists, Spec. Publ. 7*, 55–76.

Silverman, M. P. and Munoz, E. F. (1970). Fungal attack on rocks, solubilization and altered infrared spectra. *Science*, 169, 985–7.

Simonett, D. S. (1960). Soil genesis on basalt in north Queensland. *Trans. 7th int. Congr. Soil Sci.*, 4, 238–43.

— (1967). Landslide distribution and earthquakes in the Bewani and Torricelli Mountains, New Guinea. In Jennings, J. N. and Mabbutt, J. A. (eds.), *Landform Studies from Australia and New Guinea*. A.N.U. Press, Canberra, 64–84.

Simons, D. B. (1969). Open channel flow. In Chorley, R. J. (ed.), *Water, Earth and Man*. Methuen, London, 297–318.

— and Richardson, E. V. (1962). The effect of bed roughness on depth-discharge relations in alluvial channels. *Wat.-Supply Pap., Wash.*, 1498 E.

Simons, M. (1962). The morphological analysis of landforms: a new review of the work of Walther Penck. *Trans. Inst. Br. Geogr.*, 31, 1–14.

Skempton, A. W. (1964). Long-term stability of clay slopes. *Géotechnique*, 14, 75–102.

Smith, H.T.U. (1949). Physical effects of Pleistocene climate changes in nonglaciated areas. *Bull. geol. Soc. Am.*, 60, 1485–515.

Smith, M. K. (1974). Throughfall, stemflow and interception in pine and eucalypt forest. *Aust. For.*, 36, 190–7.

Smith, N. D. (1971). Transverse bars and braiding in the Lower Platte River, Nebraska. *Bull. geol. Soc. Am.*, 82, 3407–19.

Smith, T. R. and Bretherton, F. P. (1972). Stability and the conservation of mass in drainage basin evolution. *Wat. Resour. Res.*, 8, 1506–29.

Speight, J. G. (1965a). Flow and channel characteristics of the Angabunga River, Papua. *J. Hydrol.*, 3, 16–36.

— (1965b). Meander spectra of the Angabunga River. *J. Hydrol.*, 3, 1–15.

— (1967a). Geomorphology of Bougainville and Buka Islands. *Land Res. Ser. CSIRO Aust.*, 26, 15–38.

— (1967b). Spectral analysis of meanders of some Australasian rivers. In Jennings, J. N. and Mabbutt, J. A. (eds.), *Landform Studies from Australia and New Guinea.* A.N.U. Press, Canberra, 48–63.

Sprunt, B. (1972). Digital simulation of drainage basin development. In Chorley, R. J. (ed.), *Spatial Analysis in Geomorphology*. Methuen, London, 371–89.

Stall, J. B. and Fok, Y. -S. (1968). Hydraulic geometry of Illinois streams. *University of Illinois, Water Resources Center Research Report*, 15.

— and Yang, C. T. (1972). Hydraulic geometry and low streamflow regimen. *University of Illinois, Water Resources Center Research Report*, 54.

Stanhill, G. (1970). The water flux in temperate forests: precipitation and evapotranspiration. In Reichle, D. E. (ed.), *Analysis of Temperate Forest Ecosystems.* Chapman and Hall, London and Springer-Verlag, Berlin, 242–56.

Stephens, C. G. (1971). Laterite and silcrete in Australia; a study of the genetic relationships of laterite and silcrete and their companion materials, and their collective significance in the formation of the weathered mantles, soils, relief and drainage of the Australian continent. *Geoderma*, 5, 5–52.

Stewart, G. A. (ed.), (1968). *Land Evaluation*. Macmillan, Melbourne.

Stoddart, D. R. (1965). Geography and the ecological approach. The ecosystem as a geographic principle and method. *Geography*, 50, 242–51.

Strahler, A. N. (1950). Equilibrium theory of erosional slopes approached by frequency distribution analysis. *Am. J. Sci.*, 248, 673–96 and 800–14.

— (1952a). Hypsometric (area-altitude) analysis of erosional topography. *Bull. geol. Soc. Am.*, 63, 1117–41.

— (1952b). Dynamic basis of geomorphology. *Bull. geol. Soc. Am.*, 63, 923–38.

— (1964). Quantitative geomorphology of drainage basins and channel networks. In Chow, Ven-Te (ed.), *Handbook of Applied Hydrology*. McGraw-Hill, New York, 4–39 to 4–76.

Strakhov, N. M. (1967). *Principles of Lithogenesis*. Oliver and Boyd, Edinburgh and Consultants Bureau, New York.

Sundborg, A. (1956). The river Klarälven, a study of fluvial processes. *Geogr. Annlr*, 38, 127–316.

Swan, S. B. St C. (1970a). Relationships between regolith, lithology and slope in a humid tropical region, Johor, Malaya. *Trans. Inst. Br. Geogr.*, 51, 189–200.

— (1970b). Landforms of Johor. Unpublished Ph.D. thesis, University of Sussex.

Sweeting, M. M. (1966). The weathering of limestones. In Dury, G. H. (ed.), *Essays in Geomorphology*. Heinemann, London, 177–210.

Takayama, S. (1965). Bedload movement in torrential mountain streams. *Tokyo Geog. Pap.,* **9**, 169–88.

Tanner, W. F. (1968). Rivers — meandering and braiding. In Fairbridge, R. W. (ed.), *Encyclopedia of Geomorphology*. Reinhold, New York, 957–63.

Taylor, G. (1958). *Sydneyside Scenery and How It Came About*. Angus and Robertson, Sydney.

Temple, P. H. (1964). Evidence of lake-level changes from the northern shoreline of Lake Victoria, Uganda. In Steel, R. W. and Mansoll Prothero, R. (eds.), *Geographers and the Tropics: Liverpool Essays*. Longmans, London, 31–56.

— (1969). Raised strandlines and shoreline evolution in the area of Lake Nabugabo, Masaka District, Uganda. In Wright, H. E. (ed.), *Quaternary Geology and Climate* (Proc. VII Congress INQUA, Vol. 16, Publs. natn. Acad. Sci., 1701), 119–29.

Tippetts — Abbett — McCarthy — Stratton and Hunting Technical Services Limited (1967). *The Jengka Triangle Report: Resources and Development Planning*. Federal Land Development Authority, Malaysia, Kuala Lumpur.

Thom, B. G. (1973). The dilemma of high interstadial sea levels during the last glaciation. *Progress in Geography*, **5**, 167–246.

Thomas, M. F. (1969). Some outstanding problems in the interpretation of the geomorphology of tropical shields. *Brit. Geomorph. Res. Gp. Occ. Pap.*, **5**, 41–9.

— (1974). *Tropical Geomorphology*. Macmillan, London.

Thornbury, W. D. (1954). *Principles of Geomorphology*. Wiley, New York.

Thornes, J. B. (1970). The hydraulic geometry of stream channels in the Xingu-Araguaia headwaters. *Geogrl J.*, **136**, 376–82.

— (1971). State, environment and attribute in scree-slope studies. *Inst. Br. Geogrs. Spec. Publs.*, **4**, 49–63.

Todd, D. K. (1964). Groundwater. In Chow, Ven-Te (ed.), *Handbook of Applied Hydrology*. McGraw-Hill, New York, 13–1 to 13–55.

Trask, P. D. (1950). Dynamics of sedimentation. In Trask, P. D. (ed.), *Applied Sedimentation*. Wiley, New York and Chapman and Hall, London, 3–40.

Tricart, J. (1957). Une lave torrentielle dans les Alpes autrichiennes. *Révue Géomorph. dyn.*, **8**, 161–5.

— (1969). Observations sur le façonnement des rapides des rivières intertropicales. *Com. Trav. Hist. Sci. Bull. Sect. Géogr.*, **71**, 289–313.

— (1960). Les types de lits fluviaux. *Inf. Géogr.*, **24**, 210–14.

— (1961). Observations sur le charriage des matériaux grossiers par les cours d'eau. *Révue Géomorph. dyn.*, **12**, 3–15.

— (1965a). *Principes et méthodes de la géomorphologie*. Masson, Paris.

— (1965b). Schéma des mécanismes de causalité en géomorphologie. *Annls Géogr.*, **74**, 322–6.

— (1965c). *Le modelé des régions chaudes, forêts et savanes.* SEDES, Paris.

— (1969). Oscillations du riveau marin et changements climatiques dans la pampa deprimida (Pampa argentine). *Bull. Assoc. franç. Etude Quat.*, **1969**, 243–68.

— and Cailleux, A. (1965). *Introduction à la geomorphologie climatique.* SEDES, Paris.

— and Vogt, H. (1967). Quelques aspects due transport des alluvions grossières et du façonnement des lits fluviaux. *Geogr. Annlr*, **49 A**, 351–66.

Tschang, Hsi-Lin (1957). Potholes in the river beds of north Taiwan. *Erdkunde*, **11**, 296–303.

— 1962). The pseudokarren and exfoliation forms on granite on Pulau Ubin, Singapore. *Z. Geomorph., N. F.*, **5**, 302–12.

Tuan, Yi-Fu (1958). The misleading antithesis of Penckian and Davisian concepts of slope retreat in waning development. *Indiana Acad. Sci. Proc.*, **67**, 212–14. (Reprinted in Schumm, S. A. and Mosley, M. P. (eds.), *Slope Morphology.* Dowden, Hutchinson Ross, Stroudsburg, Pennsylvania, 66–8.)

Turner, A. K. (1963). Infiltration, runoff and soil classifications. *J. Hydrol.*, **1**, 129–43.

Twidale, R. (1971). *Structural Landforms.* A.N.U. Press, Canberra.

Van Andel, T. H. (1951a). Transport et sédimentation dans le lit du Rhin entre Strasbourg et Wageningen (Pays-Bas). *Sédimentation et Quaternaire, France*, **1949**, 127–33.

— (1951b). Petrology of Durance River sands. *Proc. 3rd int. Congr. Sedim.*, 43–56.

Van de Graaff, R.H.M. (1963). Soils of the Hunter Valley. *Land Res. Ser. CSIRO Aust.*, **8**, 103–35.

Van Dijk, W. and Le Heux, J.W.N. (1952). Theory of parallel rectilinear slope-recession I, *Proc. ned. Akad. Wet. Ser. B*, **55**, 115–29.

Vita-Finzi, C. (1974). Chronicling soil erosion. In Warren, A. and Goldsmith, F. B. (eds.), *Conservation in Practice.* Wiley, London, 267–77.

Vogt, H. (1962). Les facteurs de la dynamique de l'Adour moyen. *Révue Géomorph. dyn.*, **13**, 49–72.

Walker, P. H. (1970). Depositional and soil history along the lower Macleay River, New South Wales. *J. geol. Soc. Aust.*, **16**, 683–96.

Walling, D. E. (1971). Sediment dynamics of small instrumented catchments in south-east Devon. *Trans. Devon Ass.*, **103**, 147–65.

— (1974). Suspended and solute yields from a small catchment prior to urbanisation. *Inst. Br. Geogrs. Spec. Publs.*, **6**, 169–92.

Ware, A. B. (1967). The Nile at the Murchison Falls: a survey by members of the Brathay Uganda Expedition. *Geogrl J.*, **133**, 481–2.

Warner, R. F. (1971). The evolution of the landscape in southern New England. *Geol. Soc. N.S.W. New England Brch; Occ. Pap.*, **1**.

— (1972). River terrace types in the coastal valleys of New South Wales. *Austr. Geogr.*, **12**, 1–22.

Wasson, R. J. (1974). Intersection point deposition on alluvial fans: an Australian example. *Geogr. Annlr*, **56A**, 83–92.

Waters, R. S. (1953). Aits and breaks of slopes on Dartmoor streams. *Geography*, **38**, 67–76.

Wensink, J. J. (1968). The Emma Range in Surinam. *Publs. Fysisch-Geogr. Lab. Univ. Amsterdam*, **13**.

Weyman, D. R. (1970). Throughflow on hillslopes and its relation to the stream hydrograph. *Bull. Ass. int. Hydrol. scient.*, **15**(3), 25–33.

White, S. E. (1949). Processes of erosion on steep slopes of Oahu, Hawaii. *Am. J. Sci.*, **247**, 168–86.

Whitmore, T. C. and Burnham, C. P. (1969). The altitudinal sequence of forests and soils on granite near Kuala Lumpur. *Malay. Nat. J.*, **22**, 99–118.

Wilhelmy, H. (1958). Umlaufseen und Dammuferseen tropischer Tieflandflüsse. *Z. Geomorph., N. F.*, **2**, 27–54.

Williams, G. P. and Guy, H. P. (1973). Erosional and depositional aspects of Hurricane Camille in Virginia, 1969. *Prof. Pap. U.S. geol. Surv.*, **804**.

Williams, M.A.J. (1968). Termites and soil development near Brocks Creek, Northern Territory. *Aust. J. Sci.*, **31**, 153–4.

Wilson, L. (1968). Slopes. In Fairbridge, R. W. (ed.), *Encyclopedia of Geomorphology*. Reinhold, New York, 1002–20.

Wischmeier, W. H. and Smith, D. D. (1958). Rainfall energy and its relationship to soil loss. *Trans. Am. geophys. Un.*, **39**, 285–91.

Woldenburg, M. J. (1966). Horton's laws justified in terms of allometric growth and steady state in open systems. *Bull. geol. Soc. Am.*, **77**, 431–4.

Woldstedt, P. (1954). Die Klimakurve des Tertiärs und Quartärs in Mitteleuropa. *Eiszeitalter und Gegenwart*, **4–5**, 5–9.

Wolman, M. G. (1955). The natural channel of Brandywine Creek, Pennsylvania. *Prof. Pap. U.S. geol. Surv.*, **271**.

— and Leopold, L. B. (1957). Flood plains. *Prof. Pap. U.S. geol. Surv.*, **282-C**, 87–107.

— and Miller, J. P. (1960). Magnitude and frequency of forces in geomorphic processes. *J. Geol.*, **68**, 54–74.

Wood, A. (1942). The development of hillside slopes. *Proc. Geol. Assn*, **53**, 128–40.

Woodyer, K. D. (1968). Bankfull frequency in rivers. *J. Hydrol.*, **6**, 114–42.

— (1970). The terraces of the Lower Colo River, New South Wales. *Search*, **1**, 164–5.

Woolnough, W. G. (1927). The duricrust of Australia. *J. Proc. R. Soc. N.S.W.*, **61**, 1–53.

Wright, L. D. and Coleman, J. M. (1973). Variations in morphology of major river deltas as functions of ocean wave and river discharge regimes. *Bull. Am. Ass. Petrol. Geol.*, **57**, 370–98.

Wurm, A. (1935). Morphologische Analyse und Experiment. Hangentwicklung, Einebnung, Piedmonttreppen. *Z. Geomorph.*, **9**, 57–87.

Yamamoto, T. and Anderson, H. W. (1967). Erodibility indices for wildland soils of Oahu, Hawaii, as related to soil forming factors. *Wat. Resour. Res.*, **3**, 785–98.

Yang, C. T. (1971a). Potential energy and stream morphology. *Wat. Resour. Res.*, **7**, 311–22.

— (1971b). On river meanders. *J. Hydrol.*, **13**, 231–53.

— (1971c). Formation of riffles and pools. *Wat. Resour. Res.*, **7**, 1567–74.

Young, A. (1960). Soil movement by denudational processes on slopes. *Nature, Lond.*, **188**, 120–2.

— (1961). Characteristic and limiting slope angles. *Z. Geomorph., N. F.*, **5**, 126–31.

— (1963). Some field observations of slope form and regolith and their relation to slope development. *Trans. Inst. Br. Geogr.*, **32**, 1–29.

— (1970). Slope form in part of the Mato Grosso, Brazil. *Geogrl J.*, **136**, 383–92.

— (1971a). Concepts of equilibrium, grade and uniformity as applied to slopes. *Geogrl J.*, **136**, 585–92.

— (1971b). Slope profile analysis: the system of best units. *Inst. Br. Geogr. Spec. Publs.*, **3**, 1–13.

— (1973). Rural land evaluation. In Dawson, J. A. and Doornkamp, J. C. (eds.), *Evaluating the Human Environment*. Arnold, London, 5–33.

Zeller, J. (1967). Flussmorphologische Studie zum Mäanderproblem. *Geogr. Helvet.*, **22**, 57–95.

Zonneveld, J.I.S. (1968). Quaternary climatic changes in the Carribbean and northern South America. *Eiszeitalter und Gegenwart*, **19**, 203–8.

— (1969). Soela's in de Corantijn (Suriname). *K. ned. Aardr. Genoots. Geogr. Tijd. B*, 45–55.

INDEX

N.S.W. = New South Wales; Vic. = Victoria; Qld = Queensland; Tas. = Tasmania; S.A. = South Australia; W.A. = Western Australia; N.Z. = New Zealand; P.N.G. = Papua New Guinea; Penn. = Pennsylvania; Calif. = California.
The names of the individual states of the United States of America and Malaysia are employed alone.
Bold face indicates page reference to figures, plates, and tables.